Biotechnology and Genetics in Fisheries and Aquaculture

Biotechnology and Genetics in Fisheries and Aquaculture

A.R. Beaumont and K. Hoare

School of Ocean Sciences
University of Wales, Bangor, UK

Blackwell
Science

Editorial Offices:
9600 Garsington Road, Oxford OX4 2DQ
Tel: 01865 776868
Blackwell Publishing, Inc., 350 Main Street, Malden, MA 02148-5018, USA
Tel: +1 781 388 8250
Iowa State Press, a Blackwell Publishing Company, 2121 State Avenue, Ames, Iowa 50014-8300, USA
Tel: +1 515 292 0140
Blackwell Publishing Asia Pty Ltd, 550 Swanston Street, Carlton South, Victoria 3053, Australia
Tel: +61 (0)3 9347 0300
Blackwell Wissenschafts Verlag, Kurfürstendamm 57, 10707 Berlin, Germany
Tel: +49 (0)30 32 79 060

First published 2003 by Blackwell Science Ltd

Library of Congress
Cataloging-in-Publication Data
is available

0-632-05515-4

A catalogue record for this title is available from the British Library

Set in Times and produced by
Gray Publishing, Tunbridge Wells, Kent
Printed and bound in Great Britain by
MPG Books, Bodmin, Cornwall

For further information on
Blackwell Science, visit our website:
www.blackwell-science.com

Contents

List of boxes

Preface

The idea for this book was spawned by marine biology graduates at the School of Ocean Sciences, University of Wales Bangor, who proposed that A.R.B.'s Genetics in Aquaculture lecture course notes be packaged into a handbook. What seemed a relatively simple task has, of course, expanded into a larger enterprise. As with all spawnings in aquaculture, there are bound to be some instances of less than perfect development and for this we accept full responsibility. However, we hope that we have produced an introductory-level text which can explain to both students and professionals in fisheries and aquaculture what the new technologies in molecular biology and genetics have to offer.

The authors would like to thank the following for granting permission to use material in this book: Drs Ann Wood, Karen Abey, Halina Sobolewska, Shelagh Malham and Craig Wilding, and Chris Beveridge; Professors John Avise and Steve Karl; copyright holders *The Journal of Shellfish Research*, Cambridge University Press, The American Association for the Advancement of Science, The National Research Council of Canada Research Press, Elsevier Science and The Washington Sea Grant Program, University of Washington.

We are grateful to David Roberts and Geraint Williams of the School of Ocean Sciences, University of Wales Bangor for converting our sketches into publishable illustrations. Finally, we thank Nigel Balmforth of Blackwell Science for his encouragement and patience during the preparation of this book.

A.R. Beaumont & K. Hoare

'How inappropriate to call this planet Earth,
when it is clearly Ocean.'

Arthur C. Clarke

Chapter 1
What is Genetic Variation?

Have you ever seen someone who looks and sounds exactly like you? Have you ever seen your 'spitting image'? Unless you are one of a pair of monozygotic twins (twins produced by the division of a single egg) you will not have done so. It is commonly accepted that all humans are different – indeed all humans there have ever been were unique and were different from all humans living today. If this is true for *Homo sapiens*, is it also true for other sexually reproducing organisms? The answer is yes. Every salmon (*Salmo salar*) is different from every other salmon that has ever lived. Every mussel (*Mytilus edulis*) is different from every other mussel that has ever lived. This uniqueness of individuals within a species is the consequence of two factors: one is deoxyribose nucleic acid (DNA) and the other is sexual reproduction. These two factors produce and maintain the genetic diversity within a species, and an understanding of this is fundamental to our ability to sustainably exploit species of plants and animals.

Deoxyribose nucleic acid: DNA

The discovery by Watson and Crick of the structure of DNA in 1953 was a landmark in our understanding of how genetic information passes from generation to generation. In the half century since then, the fields of molecular biology and genetics have become inextricably linked and developments, particularly over the past 25 years, have opened up the potential of DNA biotechnology.

The structure of DNA enables it to carry the information for a cell to reproduce itself. It is a polymeric molecule, that is, made up of a chain of subunits, consisting of chains of nucleotide monomers. Each nucleotide contains a base, along with a sugar (deoxyribose) and a phosphate group (Fig. 1.1). There are four individual bases, adenine, guanine, thymine and cytosine and they are usually referred to by their first letter abbreviations, A, G, T and C. Two of the bases, A and G, have a double-ring structure and are known as purines. The other two bases, T and C, are pyrimidines with a single carbon–nitrogen ring.

Each nucleotide is a single unit that joins with neighbouring nucleotides in a linear fashion to make up a polynucleotide chain. Particular carbon atoms in the 5-carbon structure of deoxyribose are referred to by numbers, 1' (one prime) to 5'. The link between nucleotides is formed when the 5' of one bonds to the 3' of the next via a phosphodiester bond (Fig. 1.1). It is the sequence of the four bases in a polynucleotide chain which acts as the code for genetic information.

The complete DNA molecule actually consists of two polynucleotide chains, or *strands*, wrapped around each other in the form of a double helix. The sugar + phosphate backbones are at the outside of the molecule while the bases point towards the

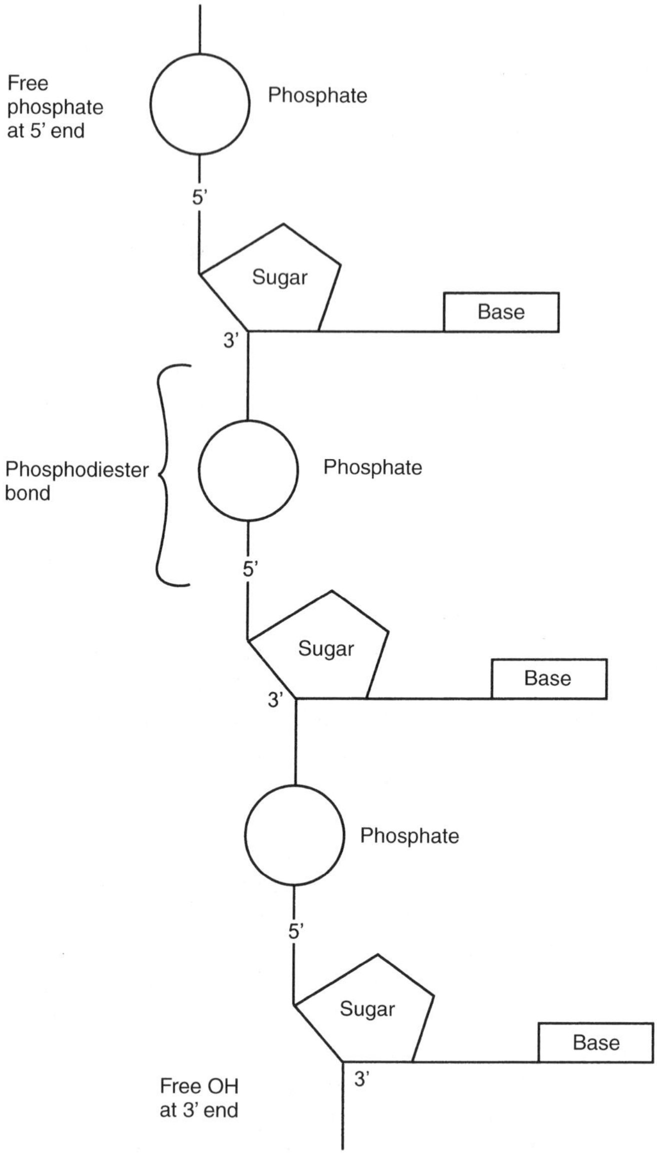

Fig. 1.1 The structure of DNA. Each nucleotide consists of a sugar, a phosphate and a base. Nucleotides are joined by a phosphodiester bond between the 5' of one ribose sugar and the 3' of the next. The chain of nucleotides therefore has a 3' and a 5' end.

middle of the structure; the two strands of the molecule run in opposite directions (Fig. 1.2).

The functional beauty of the DNA molecule is a result of complementary base pairing where G can only bond with C, and A can only bond with T, at the middle of the molecule. It means that the two strands are complementary such that the base

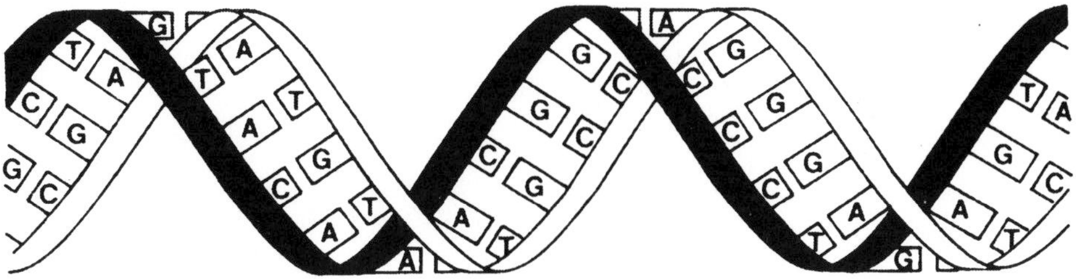

Fig. 1.2 The structure of DNA. Two polynucleotide strands are wrapped around each other in the form of a double helix. Complementary base pairing occurs between the two strands such that guanine (G) always bonds with cytosine (C) and adenine (A) always bonds with thymine (T). (Modified from Utter *et al.* (1987) Interpreting genetic variation detected by electrophoresis. In: *Population Genetics and Fishery Management* (eds. N. Ryman & F. Utter), pp. 21–45, with permission from Washington Sea Grant Program, University of Washington.)

sequence of one strand predicts and determines the base sequence of the other strand. Because one strand predicts the other it can be used to replicate the sequence.

The replication process produces daughter molecules, each of which has one parental strand and one copied strand. This is called semi-conservative replication. Replication of DNA takes place every time a cell divides. The cell's entire DNA is progressively unwound revealing short single-stranded regions which can be copied by DNA polymerase enzymes. Unwinding does not begin at the ends of the molecule, but at points called replication origins, and it then proceeds from these points along the DNA. The new strands of DNA being synthesised during replication are always synthesised in the 5' to 3' direction. This means that as the original strands separate, one new strand can be continuously synthesised against its copy strand (the leading strand) while the other has to be synthesised intermittently in short lengths as enough copy strand (the lagging strand) becomes available (Fig. 1.3).

Considering the enormous numbers of bases and coded information in the DNA of a cell, replication needs to be extremely accurate. Even a very small incidence of mistakes in copying would result in the loss of important genetic information within a few cell divisions. However, during the replication process various proofreading activities take place and almost all errors are corrected by removing the incorrect base and inserting the correct one. In spite of proofreading, a few errors are inevitable when such high numbers of bases are to be copied and it is estimated that about one in every 3 billion bases is incorrectly inserted. Such errors are called point mutations and they can also be induced by certain chemicals and radioactivity. Although there are very few of them, they are nevertheless the fundamental source of variation which fuels the process of evolution. Without such errors, no genetic change at the DNA level would take place, but with too many errors daughter cells would too often be non-viable and the organism carrying that DNA would soon become extinct.

Functional sequences only represent a small fraction of the total genome, for example around 3% in humans. The rest is made up of what has been called 'junk

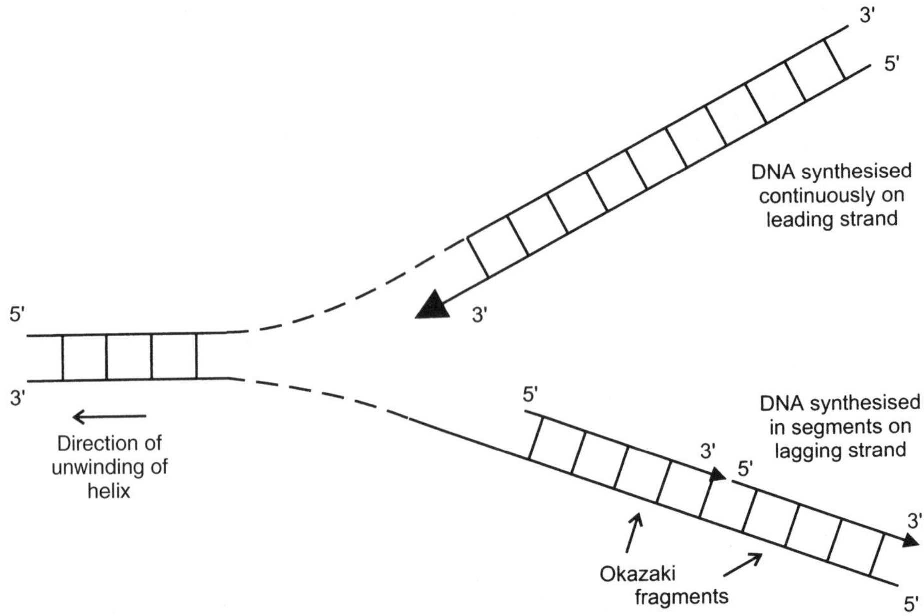

Fig. 1.3 Replication of DNA. As the DNA double helix unwinds, new DNA is synthesised continuously in a 5' to 3' direction on the leading strand and in 5' to 3' directed segments (Okazaki fragments) on the lagging strand.

DNA'. Whether all of it is really 'junk' is not known, but it is possible that much of it will have some, as yet undiscovered, function in the organism. Some of this junk DNA consists of pseudogenes, genes that for some reason or another have become non-functional. Yet other parts of non-coding DNA consist of dispersed or clustered repeated sequences of varying length, from one base pair (bp) to thousands of bases (kilobases, kb) in length. The dispersed repeated sequences occur as copies spread across the genome and can be categorised as long or short interspersed nuclear elements (LINE or SINE), long terminal repeats (LTR) and DNA transposons. The clustered repeated sequences, where the repeated sequence occurs in tandem copies, are classed as satellites, minisatellites or microsatellites depending on the length of the repeat unit, and these have turned out to be useful genetic markers, as will be explained in later chapters. Between them, these repeated elements can constitute up to 40% of the genome.

A gene is a unit of information which is held as a code in a discreet segment of DNA. This code specifies the amino acid sequence of a protein. Scientists were surprised to discover quite early on that the sequence information for a single gene was not continuous along the DNA, but was interspersed with pieces of non-coding sequence. The coding parts of a gene sequence are exons, and the non-coding parts are introns (Fig. 1.4). Before a gene can be expressed, the DNA that encodes it has to be transcribed into RNA.

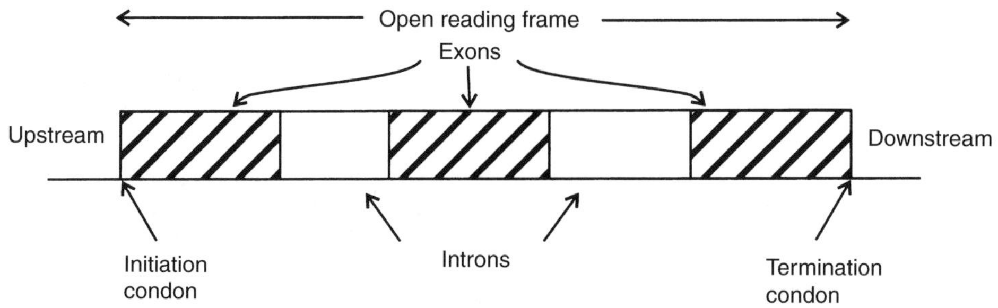

Fig. 1.4 Generalised structure of a gene. The open reading frame (ORF) for a gene begins with an upstream initiation codon and ends with a downstream termination codon. Many genes have a region, or regions, of non-coding DNA within them. These introns are spliced out of the messenger RNA during transcription so that only the codons within the exons are translated into amino acids.

Ribose nucleic acid: RNA

The structure of ribose nucleic acid (RNA) is similar to that of DNA (deoxyribose nucleic acid) except that (a) the sugar is ribose instead of deoxyribose, (b) in the place of thymine, a similarly structured base called uracil (U) is present and (c) the molecule consists of only a single polynucleotide strand. RNA molecules are produced by the process of transcription of the linear sequence of bases in DNA and are then used in the translation of that sequence into a chain of amino acids that go to make up a protein. The type of RNA transcribed from the sequence is called messenger RNA (mRNA) and translation of this sequence into a string of amino acids is undertaken by ribosomal RNA (rRNA) and transfer RNA (tRNA) molecules.

During transcription of the DNA, an RNA copy is made of one of the strands of DNA (Fig. 1.5). The two strands of DNA are called the template strand and the non-template strand. Other names for the non-template strand are the sense (+) strand or the coding strand. The RNA is synthesised by RNA polymerase enzymes using the template strand and is therefore a copy of the non-template (sense or coding) strand of DNA. Because this RNA is a direct copy of the DNA it will contain both the coding (exons) and the non-coding sequences (introns) of the gene. Introns are removed from this pre-messenger RNA and the subsequent molecule is the final mRNA. The mRNA molecules are transported from the nucleus into the cytoplasm where the message is translated into a sequence of amino acids by rRNA in bodies known as ribosomes. Amino acids are brought to the ribosomes by tRNA molecules, each specifying a particular amino acid (Fig. 1.5), and synthesised, in the presence of rRNA, into a linear sequence.

The detailed mechanics and biochemistry of the processes of transcription and translation are outside the scope of this book, but can be found in most standard genetic texts. Some modern texts use the term 'gene expression' to encompass both of these processes and their various controlling steps. For the purposes of this book, the reader need only appreciate the key concept that a sequence of bases in DNA

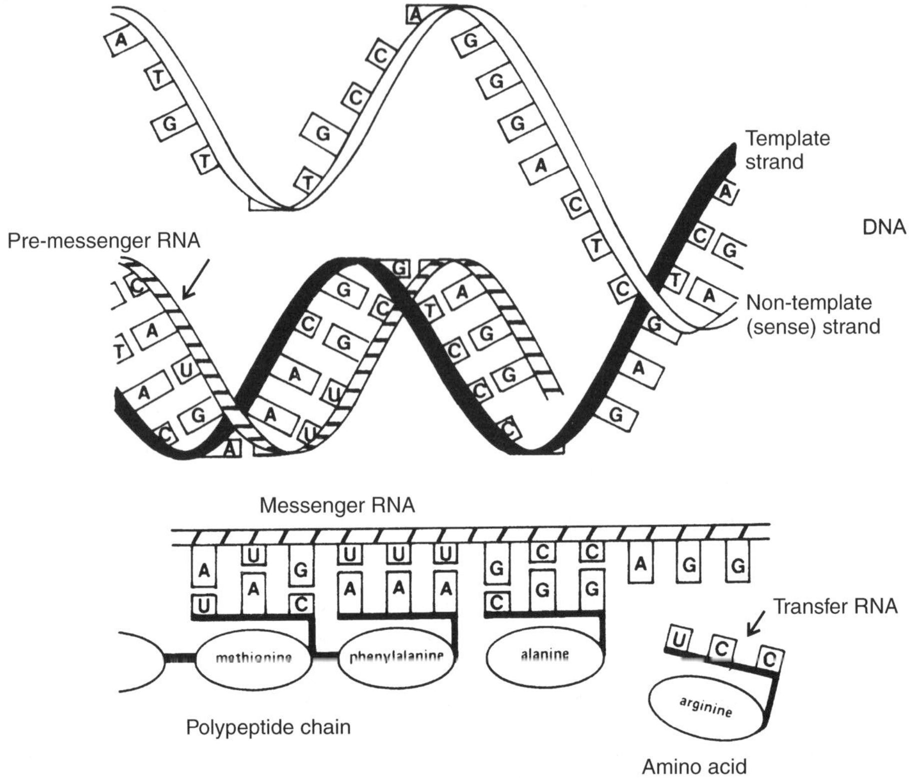

Fig. 1.5 Transcription and translation of DNA. Introns are removed from pre-messenger RNA before translation takes place. The polypeptide chain is formed from amino acids coded for by the messenger RNA and brought together by transfer RNA. (Modified from Utter *et al.* (1987) Interpreting genetic variation detected by electrophoresis. In: *Population Genetics and Fishery Management* (eds N. Ryman & F. Utter), pp. 21–45, with permission from Washington Sea Grant Program, University of Washington.)

leads, by a direct copying process involving RNA, to the production of a sequence of amino acids, the building blocks of proteins. This is what has been called the central dogma: information is transferred from DNA to RNA to protein.

What is the genetic code?

How are the four bases (A, C, G and T) in DNA organised to provide an unambiguous code for the 20 amino acids present in proteins? The 'words' of the code consist of three bases. There are $4^3 = 64$ possible combinations of the four bases into a triplet code and it is these 64 triplet codons which define the 20 amino acids. Because there are more than 20 codons, the genetic code has some redundancy – most amino acids are coded for by more than one codon. The codons are written using the symbol U,

Table 1.1 The genetic code showing the amino acids coded by the 64 triplet combinations of the four bases. The bases down the left hand side represent the first position in the reading frame, the bases along the top indicate the second position and the bases down the right-hand side show the third position

	2nd base				
1st base	U	C	A	G	3rd base
U	Phe	Ser	Tyr	Cys	U
	Phe	Ser	Tyr	Cys	C
	Leu	Ser	Stop	Stop	A
	Leu	Ser	Stop	Trp	G
C	Leu	Pro	His	Arg	U
	Leu	Pro	His	Arg	C
	Leu	Pro	Gln	Arg	A
	Leu	Pro	Gln	Arg	G
A	Ile	Thr	Asn	Ser	U
	Ile	Thr	Asn	Ser	C
	Ile	Thr	Lys	Arg	A
	Met	Thr	Lys	Arg	G
G	Val	Ala	Asp	Gly	U
	Val	Ala	Asp	Gly	C
	Val	Ala	Glu	Gly	A
	Val	Ala	Glu	Gly	G

Abbreviations for amino acids: Alanine (Ala), Arginine (Arg), Asparagine (Asn), Aspartic acid (Asp), Cysteine (Cys), Glutamic acid (Glu), Glutamine (Gln), Glycine (Gly), Histidine (His), Isoleucine (Ile), Leucine (Leu), Lysine (Lys), Methionine (Met), Phenylanaline (Phe), Proline (Pro), Serine (Ser), Threonine (Thr), Tryptophan (Trp), Tyrosine (Tyr), Valine (Val).

for uracil (in mRNA), rather than T, for thymine (in DNA). Three codons (UAA, UAG and UGA) do not encode amino acids but act as signals for protein synthesis to stop and are called termination codons or stop codons. The triplet AUG codes for methionine (formyl methionine in bacteria and mitochondria) and is the signal for protein synthesis to start. It is thus the initiation codon which sets the reading frame. The amino acid sequence of all proteins therefore starts with methionine but this is sometimes removed later. Details of the amino acids encoded by the various codons are given in Table 1.1. Note that the redundancy of the code is not random. In particular, the first two bases of the codons for an amino acid are usually the same. It is generally only the third base which varies.

Protein structure

Proteins have many tasks. Some form the structure of tissues, others – the enzymes – act as extremely specific catalysts of biochemical reactions, and yet other proteins, such as hormones, have a regulatory function. By their very nature proteins are bound to be highly complex molecules, but it is possible to categorise their structure into four basic levels. The primary structure of a protein is the linear sequence of the chain of amino acids (the polypeptide chain) and this, as we have seen already, is directly related to the sequence of bases in the DNA which codes for it. Although most amino acids are pH neutral, two are negatively charged and two positively charged. In addition, some are hydrophilic (attracted to water) and others hydrophobic (repelled by water). Thus, protein secondary structure is based on characteristic patterns produced by the properties and interactions of particular types of amino acids within the chain. One such secondary structure is an alpha-helix, another is a pleated sheet. The tertiary structure is dependent on how these secondary structures become folded in three dimensions. Therefore the DNA code, through the linear relationship of the various amino acids, dictates both the secondary and tertiary structures. This is an important point because it reveals that point mutations in the DNA coding for a particular protein can have far-reaching consequences on the final size, shape and overall charge of that protein.

Many proteins are composed of two or more polypeptide chains (subunits) and the subunits making up a protein may be identical or they may be different. The generic name for proteins with more than a single subunit is oligomers. This is the level of quaternary structure of proteins and it enables larger proteins to be produced without requiring a very long gene sequence in the DNA. It also allows greater functionality in proteins by combining different activities within a single molecule. Proteins with a single subunit are called monomers, those with two subunits are dimers and those with four are tetramers. For example glucose-phosphate-isomerase, an enzyme involved in the production of energy from the breakdown of carbohydrates, is a dimer, with two subunits coded by the same gene, while haemoglobin, which carries oxygen around in the blood, is a tetrameric molecule consisting of two alpha-globin and two beta-globin chains each coded by different genes.

So what about chromosomes?

In fish and shellfish, as with all other eukaryote organisms, the DNA molecules in the nucleus are combined with proteins, mainly histones, to make chromosomes. Each chromosome represents a single DNA molecule. Chromosomes are usually only clearly visible and identifiable when cells are dividing, at which time the chromosomes have already divided into daughter chromatids. However, the daughter chromatids retain connection to each other at a position called the centromere, or primary constriction, and this is the last part of the chromosome to divide. The position of the centromere on the chromosome can be central (metacentric), between the centre and one end (submetacentric), very close to one end (acrocentric), or terminal (telocentric). The number of chromosomes, their lengths and the positions of their centromeres are unique to each species and these characters are used as descriptors for the species karyotype (Fig. 1.6). Chromosomes themselves mutate and evolve (Box 1.1) and before the advent of allozyme markers some geneticists spent

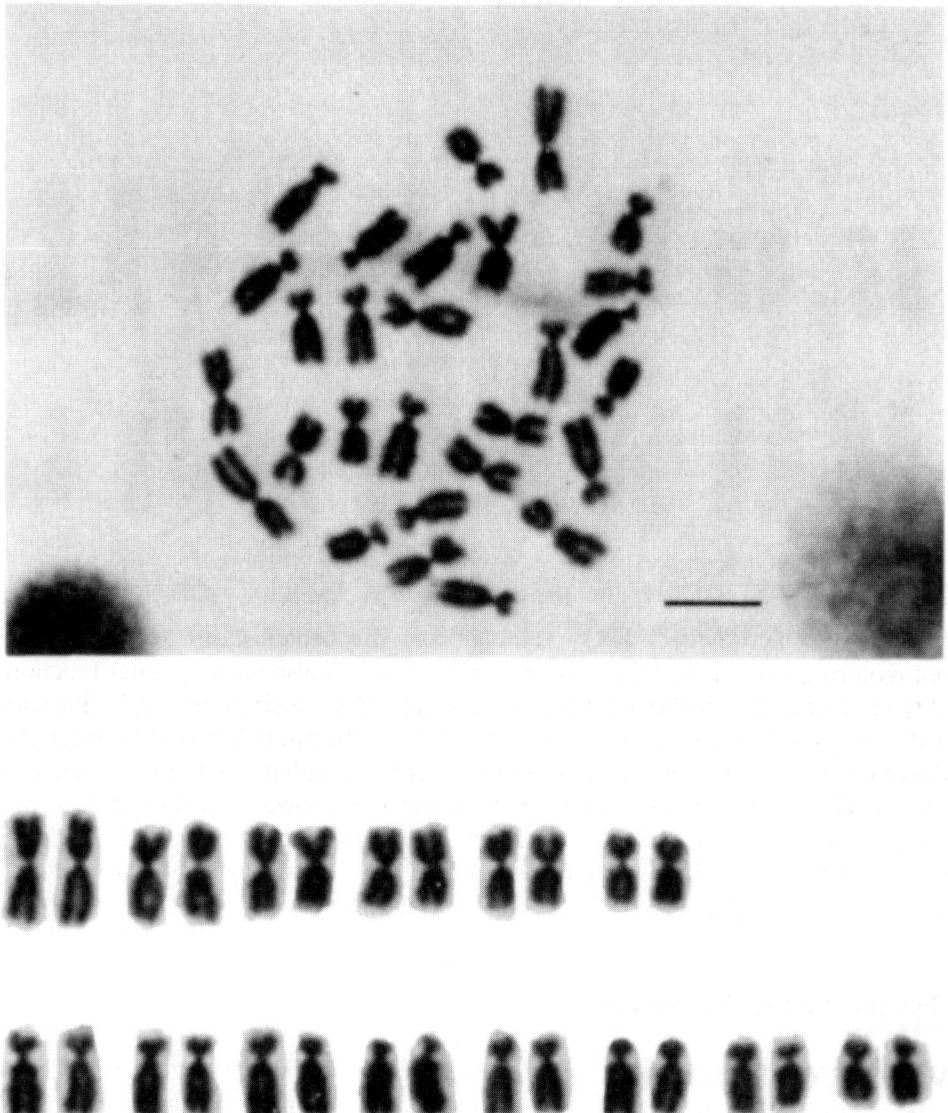

Fig. 1.6 Metaphase chromosome spread and karyotype of *Mytilus edulis*. There are six metacentric and eight submetacentric pairs of chromosomes. Haploid number $N = 14$, diploid number $2N = 28$. Scale bar = 5 μm. (Reproduced with permission from Dixon, D.R. & Flavell, N. (1986) A comparative study of the chromosomes of *Mytilus edulis* and *Mytilus galloprovincialis*. *Journal of the Marine Biological Association, UK*, **66**, 219–228, Cambridge University Press.)

much of their time squinting down microscopes following the inheritance of chromosomal rearrangements. Nowadays, chromosomal variation is assessed for aquaculture and fisheries purposes, mainly in relation to interspecies hybridisations.

Box 1.1 Genetic variation at the level of the chromosomes

Although chromosome variations are no longer used as markers in population genetic studies, they play an important role in evolution.

Most chromosome rearrangements arise, as do point mutations, as a result of mistakes during the replication of the DNA molecule. Such rearrangements, however, involve long segments of DNA, rather than single bases.

Chromosome deletions occur when the DNA strand breaks but fails to mend. Fragments of chromosome produced in this way that do not contain a centromere (acentric fragments) will be lost during subsequent cell divisions.

Chromosome duplications provide an extra copy of a block of DNA that may contain complete gene sequences. As might be expected, duplications are less harmful than deletions and when duplications contain complete gene sequences natural selection can operate independently on both the new and the old sequences to produce divergent roles for the genes. This is the principal process for the evolution of new genes.

Sometimes a fragment of one chromosome can become exchanged with a fragment of another non-homologous chromosome and such an exchange is called chromosomal translocation.

A chromosomal inversion is where a fragment of chromosome breaks off and reattaches to its original position in reversed orientation. The inverted fragment may have contained the centromere (pericentric inversion) or it may not (paracentric inversion).

There is one further type of chromosomal rearrangement which needs a mention. This involves fusion or fission of the centromere. Two telocentric or acrocentric chromosomes may fuse at their centromeres to produce a single bi-armed chromosome and this is called a Robertsonian translocation. Alternatively, a bi-armed chromosome can break at the centromere to produce two telocentric chromosomes. These types of chromosomal rearrangements may explain much of the variation in chromosome number between species.

Before allozyme electrophoresis provided geneticists with access to individual genes for study, many geneticists spent their time looking at the structure of chromosomes during meiosis. The structure of the paired chromosomes (bivalents, Fig. 1.7) observed during meiosis reflects chromosomal rearrangements – chromosomes with translocations can only pair up by forming chains or rings, inversions produce loops in the bivalent, etc. Banding patterns on chromosomes shown up by particular stains (e.g. G-banding, produced by Giemsa stain) can also be used to characterise chromosomes and their rearrangements. Although used in early studies of heritable variation, chromosomal rearrangements are usually deleterious and often result in a non-viable gamete.

Chromosomal rearrangements may explain much of the variation in chromosome number between species and examination of karyotypes between closely related species is important when considering artificial hybridisations between them.

Polyploidy is a condition where individuals have more than two copies of each chromosome. For example, triploids have three sets of chromosomes and tetraploids have four. Polyploidy occurs naturally in some plants (e.g. wheat, which is hexaploid) and a tetraploidisation event has occurred in the recent evolutionary history of salmonids. Polyploidy can be artificially induced in normally diploid species for aquacultural purposes as will be seen in Chapter 6.

Almost all fish and shellfish are diploids. Diploids have two complete sets of DNA instructions, so that each chromosome is just one of a homologous pair. Normal cell division – mitosis – combines division with replication because prior to cell division each chromosome replicates (into two chromatids) and one copy passes into each daughter cell. Diploidy is thus maintained. During the process of sexual reproduction a specialised cell division called meiosis takes place which produces daughter cells, the gametes, which have only a single set of chromosomes. Gametes are therefore haploid. We will be looking in more detail at the process of meiosis later in this chapter.

If an organism inherits the same version of a gene from both parents, it is said to be homozygous. If the two versions are different, the organism is heterozygous. Each version of a particular gene is called an allele; the two alleles possessed by a diploid organism at each locus (position on the chromosome, plural loci) make up its genotype for that locus.

In many organisms there is a special pair of chromosomes which defines the sex of their carriers. For example in the XX–XY system present in humans and some fish, females have a pair of identical sex chromosomes (the X chromosomes) while males have one X chromosome and a reduced size Y chromosome. The other chromosome pairs are called autosomes. In many shellfish there are no identifiable sex chromosomes and, in the case of certain molluscs such as oysters, individuals may even change their sex during their lives.

How does sexual reproduction produce variation?

The key feature of sexual reproduction is the production of haploid gametes through the process of meiosis and the uniting of these gametes to produce a new diploid generation. The process of meiosis shuffles the genetic material in such a way that none of the haploid chromosome sets in the gametes are identical to either of the haploid sets present in the parent from which they are derived. To see how this happens we must look at particular stages of the process of meiosis. Here we give a brief outline of the behaviour of the chromosomes during meiosis, emphasising the genetic consequences rather than describing each stage in detail (Fig. 1.7). The process of meiosis actually consists of two cell divisions, meiosis I and meiosis II. The full details of meiosis are given in all standard genetic texts.

Meiosis I begins long before the chromosomes become clearly visible. The chromosomes are initially very thin and uncontracted but become progressively more con-

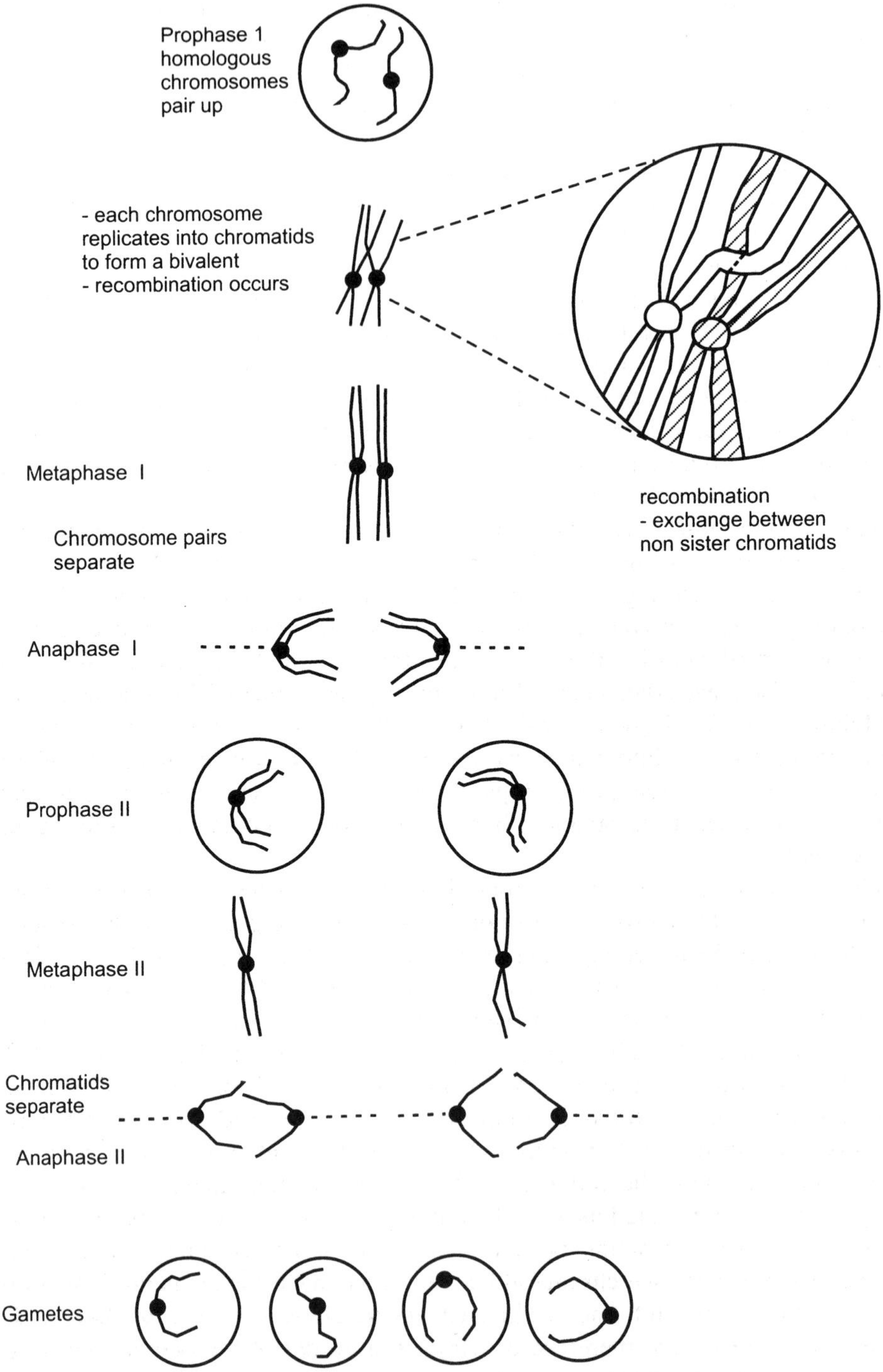

Fig. 1.7 The process of meiosis. Recombination takes place during prophase of meiosis I.

tracted and more visible during the prophase stage. During this stage, homologous pairs of chromosomes come to lie closely together and at the same time each chromosome in each pair divides into chromatids that remain attached to one another at the centromere. So each pair of chromosomes consists of four chromatids. Such pairs of chromosomes at this stage are called bivalents. It is during this time, while the chromosome pairs are adhered closely together, that the process of recombination or crossing-over occurs. Recombination involves the interaction between two ordinary (double-stranded) DNA molecules. The effect is that both molecules break, but the ends rejoin to the 'wrong' molecule. From the genetic point of view, each chromatid in a bivalent can be considered to be effectively a single DNA molecule and the recombination interaction takes place between chromatids that derive from different chromosomes of the pair – the non-sister chromatids (as opposed to the sister chromatids which are derived from an individual chromosome). These recombinations take place in every chromosome pair, usually one per chromosome arm, but sometimes more than one. The place where a recombination event is located on a bivalent is visible as a chiasma.

At metaphase of meiosis I the bivalents lie across the equator of the spindle with their centromeres attached to the arms of the spindle. Meiosis I is a reduction division (Fig. 1.7). During anaphase, each pair of chromosomes is separated so that one of the pair goes into one daughter cell and the other into the other daughter cell. So at the start of meiosis I the cell contains four copies of the genetic information, but after division each daughter cell contains only two copies of the genetic material.

Meiosis II is effectively a mitotic division where the two chromatids from each chromosome separate into daughter cells (Fig. 1.7). Therefore each diploid cell that enters into meiosis produces four haploid gametes. Some genetic texts suggest that a cell can be regarded as 'tetraploid' when it enters meiosis I because it has four copies of the DNA.

What is the actual effect of recombination across the genome? Take a species such as the flat oyster *Ostrea edulis*, for example, which has 10 pairs of chromosomes. In all cells of the body, including the germ cells which will undergo meiosis, one of each pair of chromosomes will have come from the female and the other from the male parent. Envisage one of the chromosomes as a linear arrangement of genes along a single molecule of DNA. The other chromosome of the homologous pair will also consist of that same linear arrangement of genes along its length but will have come from a different parent. It has the same genes, in the same order, but has a different ancestry. That ancestry will have provided it with different variations at many of its genes compared with the other chromosome of the pair. In early meiosis each chromosome has replicated itself so there are two DNA copies (chromatids) of each chromosome. Recombination occurs between non-sister chromatids such that a stretch of DNA from one chromatid becomes exchanged for the equivalent stretch from the other chromatid. The resulting chromatid DNA molecules that have undergone recombination are therefore different from either of the parental ones. Any chromatids which have not been involved in a recombination event, of course, remain unaltered.

Now note that this process is taking place in all of the 10 pairs of chromosomes in that germ cell during that division. Then consider that this is just one germ cell among the millions of germ cells in the gonad of the oyster. All the other germ cells are also undergoing a meiotic division during which recombination is taking place in all the pairs of chromosomes. The precise position along the DNA molecules (chromatids) at which recombination events take place is (to some extent) random and will generally be different in each dividing germ cell. So it is easy to understand why the 10 DNA molecules (chromosomes) in an oyster gamete are going to be different from any of the 10 parental DNA molecules (chromosomes) that were present in the germ cell before meiosis. It can also be understood why the genetic make up (the 10 DNA molecules) of every gamete is likely to be different from every other gamete.

It can be seen that very extensive shuffling of the genome is achieved by recombination. However, this is not the only reshuffling that takes place during meiosis. Consider the 10 pairs of chromosomes in the germ cell, with one from each pair derived from the male parent of the oyster, and the other from the female parent. These can be indicated as M1, M2, M3 up to M10 (for the male derived chromosomes) and F1, F2, F3 up to F10 (for the female). Each of the daughter cells following meiosis I will contain 10 chromosomes, but they will be a random mixture of F and M chromosomes as illustrated in Figure 1.8. For 10 chromosomes there are $2^{10} = 1024$ possible combinations. This is called independent assortment of chromosomes.

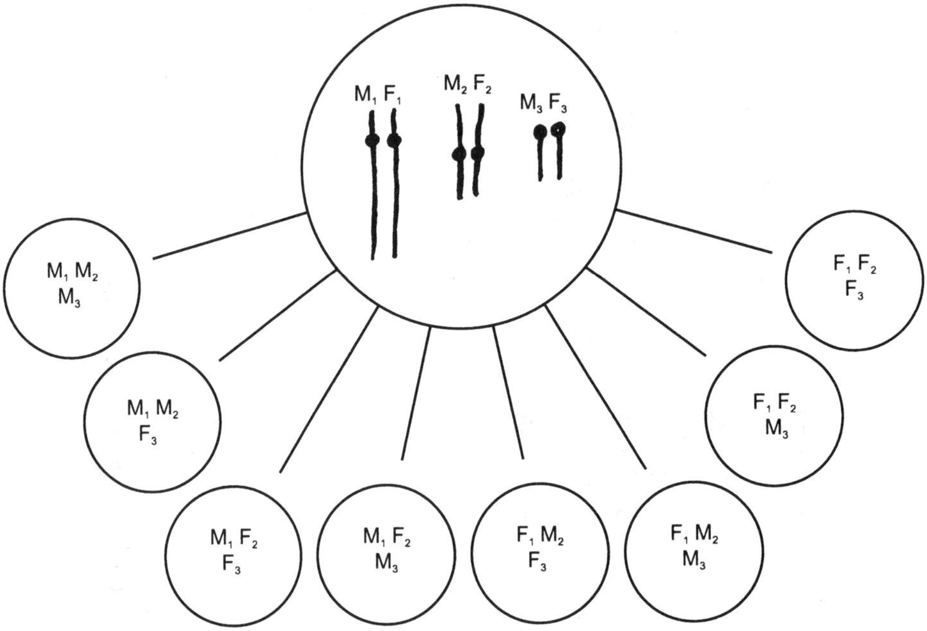

Fig. 1.8 How independent assortment of chromosomes at the end of meiosis I creates variation. Three pairs of chromosomes are indicated: M1, M2 and M3 are the male parental chromosomes and F1, F2 and F3 are the female parental chromosomes. Independent assortment gives eight possible combinations of chromosomes in daughter cells.

Therefore, shuffling of the genome takes place by two processes in meiosis – recombination and independent assortment – and these processes probably ensure that no two gametes are ever likely to be identical to either parental chromosome set, nor to one another.

The final part of the process of sexual reproduction, syngamy – the fusion of male and female gametes at fertilisation to form a zygote – further increases the genetic variation of offspring from their parents. Considering all these factors, it is not at all surprising that no two individuals in a sexually reproducing species are identical. The only exception is if an already fertilised egg (a zygote) divides to produce two separate cells, both of which develop independently into normal embryos. These are monozygotic twins and are effectively genetic clones of one another. The original zygote from which they arose would still have been different from any other zygote, or individual, in that species.

Although the chromosomal behaviour during meiosis is the same in both males and females, there is an important difference in the production of spermatozoa and eggs (ova). In males, four spermatozoa are produced from each germ cell. In females, only one egg (ovum) is produced from each germ cell or primary oocyte (Fig. 1.9).

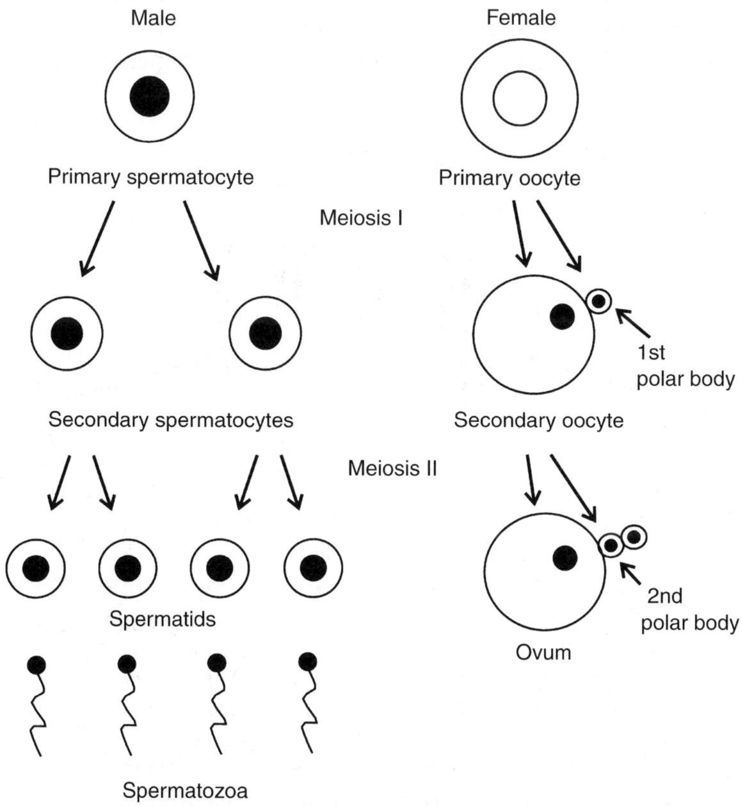

Fig. 1.9 The difference in meiotic products in males and females. In males each primary spermatocyte produces four spermatozoa while in females each primary oocyte produces a single ovum.

One of the two cells produced during meiosis I is large, and the other is very small, so small that effectively it is really just the chromosomal material with little or no cytoplasm. This small cell – the first polar body – usually does not undergo meiosis II. The large cell – the secondary oocyte – again divides unequally during meiosis II to produce the large ovum and the small second polar body.

Agricultural animals are mostly mammals or birds in which the meiotic divisions take place inside the body of the animal or inside a shell and so are not easily accessible. However, in most fish species, only meiosis I takes place before spawning and the secondary oocytes are released into the water. Meiosis II only occurs when spermatozoa have become attached. In bivalve molluscan shellfish, the oocytes are spawned at metaphase of meiosis I and further development is dependent on the attachment of spermatozoa. This feature of fish and certain shellfish enables chromosome set manipulation to be simply engineered in such species for the production of polyploids (Box 1.1). Nevertheless, in certain fish, in crustacea, and in brooding bivalves, eggs are not directly accessible during the meiotic divisions. The production of polyploids will be discussed in Chapter 6.

In addition to variation at the DNA level, and the shuffling of the genes during meiosis, genetic variation also occurs at the level of the chromosomes. Such variations are not very useful as genetic markers in aquaculture species, but chromosomal rearrangements have played an important role in evolution (Box 1.1).

Mitochondrial DNA

So far we have considered the DNA present in the nucleus, which is organised into chromosomes. There is actually more DNA in the cell – extra-chromosomal genes, contained within energy-generating organelles, mitochondria, of which there may be several hundred in each cell. In plants, the photosynthetic organelles – chloroplasts – also contain DNA.

Animal mitochondrial (mt) DNA is normally present as a circular molecule of around 16 kb in length and there are around 10 copies of the DNA in each mitochondrion in humans. Unlike the chromosomal DNA, there is no meiosis and replication appears to be a simple copying process, though the very latest research does point to there being some form of recombination during mtDNA replication.

Because there are large numbers of mitochondria in an egg, but very few in a spermatozoon, it is hardly surprising to find that the mtDNA present in a sexually reproduced offspring is usually inherited entirely from its mother. This maternal-only inheritance of mtDNA is the normal situation in almost all animals. However, one exception to this rule occurs in an important aquaculture species, the mussel *Mytilus* spp., which has a form of bi-parental inheritance of mtDNA. Females have an F type of mtDNA in every body cell while males have both the F type and an M type mtDNA in most cells of the body. The M type is highly concentrated in the male gonad and is thought to be the only mtDNA present in the spermatozoa. These M mtDNA molecules present in a spermatozoon enter the egg and, in some way which is not yet fully understood, the M type remains in the egg after fertilisation and is eliminated in individuals destined to become female, but retained and preferentially replicated in individuals destined to become males. This unusual arrangement has been named 'doubly uniparental inheritance' (DUI). DUI has recently been detected in other bivalves besides mussels and may be more widespread than currently thought.

The complete sequence of the mitochondrial genome is now known for quite a number of vertebrates and invertebrates, and the order of the genes within the circular genome is different in every phylum so far studied. Because fish mitochondrial DNA has been extensively used in phylogenetic studies, we will use this molecule as an example. The mitochondrial genome of fish contains 13 genes coding for proteins, two genes coding for ribosomal RNA (the small 12S and the large 16S rRNA), 22 genes coding for transfer RNA molecules (tRNAs) and one non-coding section of DNA which acts as the initiation site for mtDNA replication and RNA transcription. This is called the control region (Fig. 1.10).

In contrast to the nuclear genome, the mitochondrial genes of animals are very efficient and have no introns. In addition there is virtually no 'junk DNA' or repetitive sequences in the mitochondrial genome, although the control region does often vary in length due to tandem repeats. Exceptions to this general rule are the

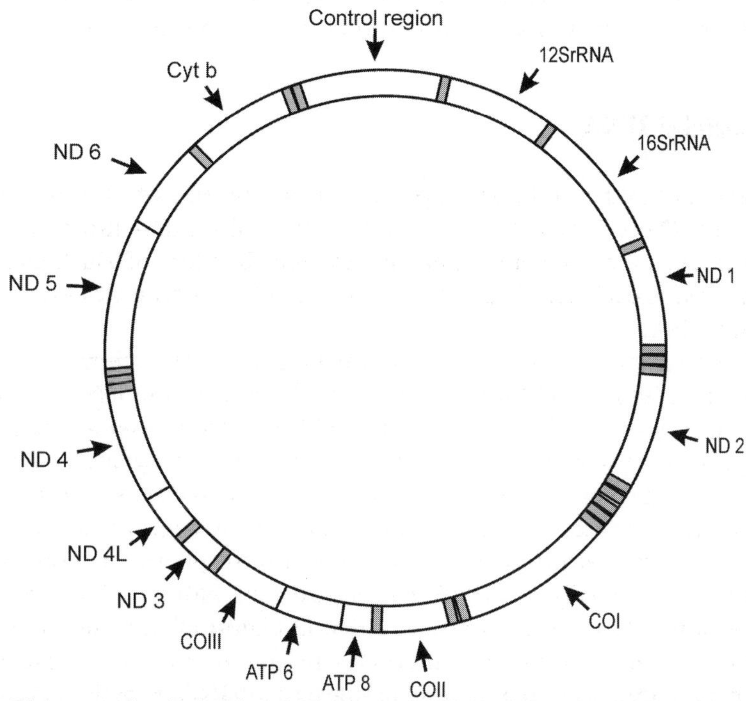

Fig. 1.10 The generalised mitochondrial genome of fish. The position of the 13 protein coding genes are indicated on the molecule. They are: seven subunits of the enzyme NADH dehydrogenase (ND 1, 2, 3, 4, 4L, 5, 6), cytochrome *b* (Cytb), three subunits of cytochrome *c* (COI, II, III) and two subunits of the enzyme adenosine triphosphate synthetase (ATP6 and ATP8). 12SrRNA = 12S ribosomal RNA, 16SrRNA = 16S ribosomal RNA. The shaded segments indicate the positions of the transfer RNA genes.

scallops, many species of which exhibit several large (up to 1.4 kb) repeated sequences within the mtDNA genome which can consequently extend to beyond 30 kb in length.

For reasons which are not fully understood, the rate of mutation in animal mtDNA is higher than in the nuclear DNA (about 5 to 10 times higher). This means that the rate of evolution is greater in mtDNA than in nuclear DNA, and this feature is of importance to us when we are looking for genetic markers which will reflect changes in the more recent past.

Further reading

Avise, J.C. (1994) *Molecular Markers, Natural History and Evolution*. Chapman & Hall, London.

Brown, T.A. (1999) *Genomes*. Bios Scientific Publishers, Oxford.

Majerus, M., Amos, W. & Hurst, G. (1996) *Evolution: the Four Billion Year War*. Longman, New York.

Turner, P.C., McClennan, A.G., Bates, A.D. & White, M.R.H. (1997) *Instant Notes in Molecular Biology*. Bios Scientific Publishers, Oxford.

Winter, P.C., Hickey, G.I. & Fletcher, H.L. (1998) *Instant Notes in Genetics*. Bios Scientific Publishers, Oxford.

Chapter 2
How Can Genetic Variation be Measured?

Genetic variation can be measured and quantified at several levels. First, the precise sequence of a length of DNA, and how it varies between individuals, can be determined. Secondly, differences between sizes of DNA fragments can be identified. At the next level we can consider protein differences that result from DNA coding sequence variation. Finally, it is sometimes possible to identify phenotypic differences that are the product of genetic variation at just one or two loci.

DNA sequence variation

The crude extraction of DNA from animal or plant tissue is a simple process which involves mechanically or chemically breaking down the insoluble cellular structures and removing them by centrifugation. Soluble cellular proteins, and the proteins which bind the DNA into the chromosomes, can then be broken down using a strong protease enzyme and removed, usually using solvents such as phenol-chloroform. The DNA is present in the water-soluble component and can then be precipitated using an alcohol. There are a number of commercial kits on the market which enable further purification of DNA. The next problem is to produce multiple copies of specific fragments of DNA and this can be done either by cloning the fragment or by the use of the polymerase chain reaction (PCR). In the process of cloning (Box 2.1), the target DNA is inserted into a vector molecule which is taken up or inserted into host

Box 2.1 Cloning

It perhaps should come as no surprise to discover that DNA is a very tough molecule and can withstand considerable stresses during its extraction. However, for accurate DNA analysis, long, unbroken molecules are required, and care is required to reduce shearing of the molecules during preparation. Once long, high molecular weight DNA molecules have been extracted and purified they can be cut into fragments using restriction endonucleases (REs) that are enzymes purified from bacteria. One class (type II) of these enzymes have the useful property of only cutting the DNA molecule at particular points in the sequence, each enzyme having its own recognition sequence of four or more bases. For example, a restriction endonuclease isolated from the bacterium *Escherichia coli*, named *Eco*RI, cuts DNA only where the hexanucleotide 5'–GAATTC–3' occurs (Table B.21). The cut is uneven, producing an overlap on each end making them 'sticky' or 'cohesive'. Other restriction endonucleases, such as *Alu*I, make blunt ended cuts (Table B2.1).

Table B2.1 Recognition sequences and type of end sequence of three commonly used restriction endonucleases

Restriction endonuclease	Recognition sequence	End sequences		Type of end
*Eco*RI	5'–GAATTC–3' 3'–CTTAAG–5'	5'–G 3'–CTTAA	AATTC–3' G–5'	Sticky
*Alu*I	5'–AGCT–3' 3'–TCGA–5'	5'–AG 3'–TC	CT–3' GA–5'	Blunt
*Hin*fI	5'–GANTC–3' 3'–CTNAG–5'	5'–G 3'–CTNA	ANTC–3' G–5'	Sticky

G = guanine, A = adenine, C = cytosine, T = thymine, N = any nucleotide.

Once DNA has been cut into fragments, the fragments can be 'pasted' into a vector using the enzyme DNA ligase. There are a number of vectors available depending on such factors as the size of the fragments to be cloned, the host organism (bacteria, yeast, plants, mammals) and whether one wishes to express (i.e. transcribe and translate) the genes on the cloned fragments. However, by far the most common cloning system for purposes relevant to aquaculture, where we tend to be probing for particular genes or marker sequences, is to use a modified form of the bacterium *Escherishia coli* as the host. There are two principle vectors used to get DNA into a bacterium, one a virus (bacteriophage or just phage) that infects the bacterium and the other a plasmid, which is a circular DNA molecule occurring as a natural inclusion in many bacteria. Some labs still prefer phages because the infectious particles naturally contain extractable single-stranded DNA that they find gives good sequencing results. However, most labs prefer to be able to sequence the DNA in both directions (which can not be achieved with only one strand) and find that plasmid vectors are less likely to 'chew up' the inserted DNA. There are many variants of plasmid and a very common and well-behaved one is pUC19 ('p' for plasmid, 'UC' for the University of California, where the plasmid was created, '19' to show that it was the nineteenth such plasmid created there).

DNA is extracted from an organism and then cut by incubation of the DNA with a restriction enzyme. The cut fragments are then mixed in solution with the enzyme DNA ligase and the vector, in this case a plasmid, which has been previously cut with the same or a compatible restriction enzyme (plasmids and other vectors have been engineered to contain a polycloning site, which contains the recognition sequences for many different restriction enzymes). Many of the DNA fragments become ligated into plasmid molecules and the vector, plus its included DNA, is then inserted into a special form of *E. coli*. This 'competent' *E. coli* takes in the vector when subjected to a shock of some kind, usually heat (Fig. B2.1a).

The plasmid vector contains the gene sequence for resistance to an antibiotic (e.g. ampicillin, chloramphenicol). The *E. coli* used has no resistance of its own.

The antibiotic is added to the agar plates so only bacterial clones that include the plasmid will grow on the plates.

The *E. coli* cells are spread very thinly over the agar plates so that each transformed cell can form a separate colony when allowed to replicate overnight at 37°C (Fig. B2.1b). As well as the bacterial multiplication, the plasmid replicates within each bacterial cell, thereby producing millions of copies of the included DNA in bacterial clones.

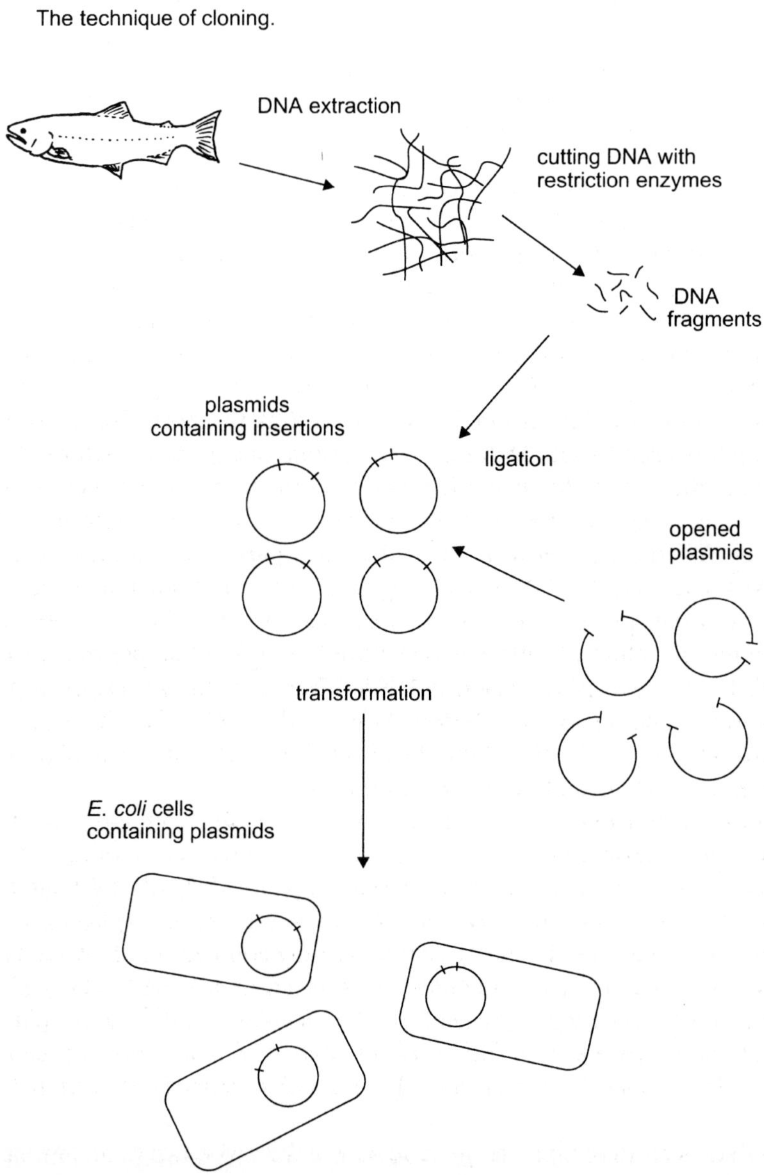

Fig. B2.1a The technique of cloning DNA fragments.

A second plasmid gene is employed to identify those colonies that contain non-recombinant plasmids, that is, bacteria which took up plasmids which had self-ligated and had no added, recombinant, DNA. The plasmid used has a gene for B-galactosidase, but the plasmid's cut site is in the middle of this gene. Therefore plasmids that have self-ligated will still have an active B-galactosidase gene, while plasmids that contain recombinant DNA will not. Using the substrate X-gal in the agar plates, which produces a blue product on reaction with B-galactosidase, enables blue-coloured colonies (non-recombinant DNA) and white colonies (recombinant DNA) to be identified.

If desired, white colonies containing recombinant DNA can be individually picked from the plates using a sterile toothpick and maintained in a 'DNA library' of clones (such DNA libraries are commercially available for some species). However, most researchers probe for the required genomic DNA sequence on the original transformed colonies. This is done by carefully laying a nylon membrane onto the plate so that some of each colony is transferred to the membrane which is then carefully peeled off. While the membrane is on the plate their relative positions are marked, for example by puncturing both with a red-hot needle, so that they can be accurately lined up again later. The membranes are treated to break down the bacterial cell walls and to separate the strands of the DNA (denaturation). The DNA is then fixed to the membrane using heat or ultraviolet light. The membranes are then probed for the sequence of interest. This is done by hybridising the plasmid DNA with a labelled probe, a short sequence of single-stranded DNA complementary to the sequence of

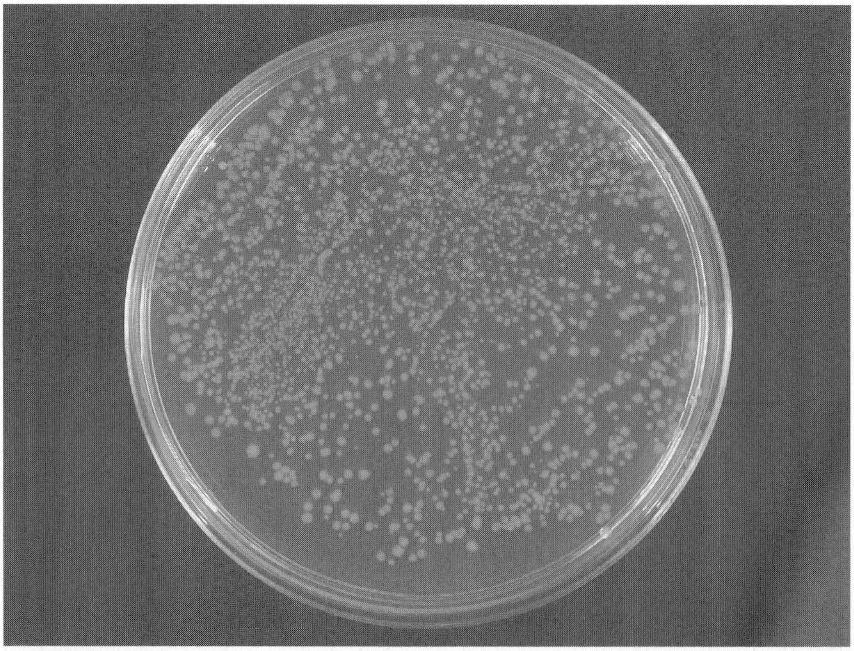

Fig. B2.1b Colonies of *E. coli* on an agar plate.

interest. Probes are generally radioactively labelled, though fluorescent labels are available. Hybridisation involves exposing the nylon membranes to the labelled probe at a temperature high enough to melt all but a very good DNA match. The probe DNA thus becomes annealed only to the target DNA, carrying its label with it. The radiolabel is visualised by autoradiography – exposure of the membranes to X-ray film (Fig. B2.1c). The needle holes on the membranes can be marked to show up on the film, so that the original agar plates with their re-grown bacterial colonies can be lined up with the autoradiograph and those clones which gave a positive radiolabel signal can be identified and isolated for sequencing or further analysis.

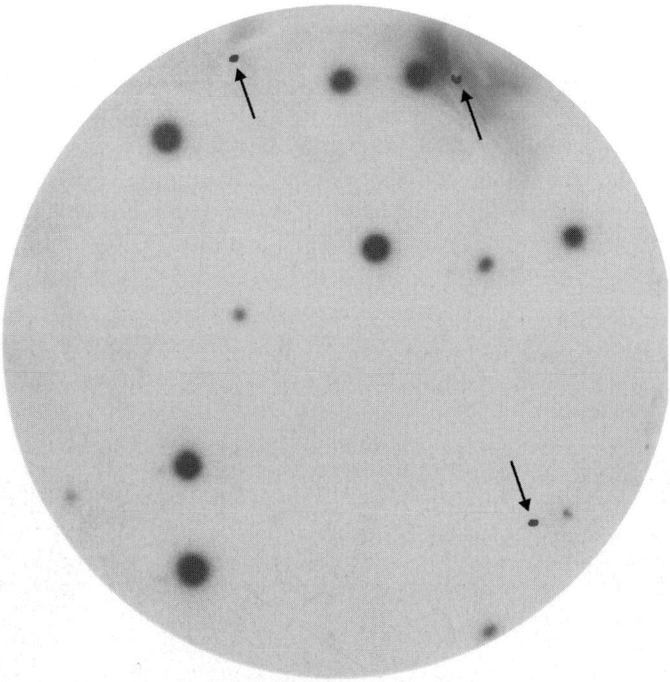

Fig. B2.1c Autoradiograph of a nylon membrane lifted from an agar plate and treated with a radiolabelled probe. Strong positive signals are evident from several clones. Arrows indicate needle marker points.

cells. Subsequent rapid replication of these host cells and the vector molecules inside them results in the production of millions of copies of the target DNA. As far as most DNA markers are concerned, cloning is usually only needed during the development phase – once the DNA sequences flanking the markers have been found from the cloned fragments, PCR can be used to produce millions of copies of the target sequence within a few hours. The PCR method relies on the fact that double-stranded DNA becomes denatured and separates into single strands when heated above 90°C. Once denatured, the temperature is lowered to a predetermined annealing temperature which allows short manufactured lengths of single-stranded DNA of known sequence (primers), designed to be complementary to the regions flanking the target DNA, to attach (anneal) to these flanking regions. Raising the temperature to 72°C in the presence of a DNA polymerase enzyme and the building blocks of DNA results in two copies of the double-stranded target DNA. Each time the cycle is repeated the number of copies is doubled and, since each cycle takes only a minute or two, millions of copies can be produced within a few hours by this method. For full details of the PCR method see Box 2.2.

Box 2.2 The polymerase chain reaction (PCR)

The PCR technique makes millions of copies of a particular target DNA sequence. The whole amplification process takes place in microtubes or in microwells in plastic plates in a small thermal-cycling machine on the bench (Fig. B2.2a). Each microtube contains a number of ingredients together with the template DNA that is to be copied from. Millions of copies of a pair of primers – short single-strand sequences of DNA each complementary to one end of the target DNA sequence – are included. A thermostable DNA polymerase enzyme (e.g. *Taq* polymerase, derived from the bacterium *Thermus aquaticus*, a resident of hot springs) is present together with the four deoxynucleotide triphosphates (dATP, dCTP, dGTP and dTTP, collectively dNTPs) in a buffer. Using *Taq* polymerase, the maximum length of the target DNA is effectively around 3–4 kb,

Fig. B2.2a A thermal cycler in which PCR is carried out. During PCR the heated lid is closed over the microtubes which are positioned in the heating block.

because longer fragments cannot be successfully amplified and inaccuracies begin to accumulate to unacceptable levels. There are special DNA polymerases available for those who need long and accurate PCR replication.

There are three stages to PCR – denaturation, primer annealing and polymerization – each one lasting only about a minute and each operating at a different temperature:

- **The denaturation step:** the contents of the microtubes are heated to above 90°C to separate the two strands of the template DNA.
- **The primer annealing step:** the temperature is decreased rapidly to a predetermined annealing temperature, usually around 55°C, to allow the primers to 'sit down', that is, to become annealed to their complementary sequences on the template DNA.
- **The polymerisation step:** The temperature is increased to 72°C, the temperature at which the *Taq* polymerase is most active, to enable the synthesis of new DNA in the 3' direction away from the primers.

These steps are then repeated:

- **The denaturation step (2):** the mixture is again heated to about 94°C to denature all the newly built molecules, and any other parts of the template DNA which have become annealed by chance.
- **The primer annealing step (2):** primers anneal as in the first cycle, but this time some will anneal to the newly manufactured strands of DNA.
- **The polymerisation step (2):** new synthesis of molecules takes place and results in some molecules which have one strand of the precise length defined by the primer sequences at each end.
- **The denaturation step (3):** the strands of DNA are again separated ready for the annealing step.

From this point on, the number of newly synthesised molecules of the precise length specified by the two primers increases exponentially at each new cycle. Usually 20 to 40 cycles are used and the resulting PCR product should consist of a very high copy number of the target sequence together with a small amount of original and fragmented DNA (Fig. B2.2b).

The above describes the theory. In practice, the PCR method is extremely sensitive to small variables. For each pair of primers, there is an optimum annealing temperature and optimum Mg^{2+} concentration. The slightest contamination of the template DNA with proteins or other material can often inhibit PCR amplification. *Taq* polymerase and buffers from different manufacturers have slightly different characteristics and may require re-optimisation. And, of course, while temperatures are changing between steps, all the ingredients are free to interact in the most unpredictable way. In spite of these considerations, the PCR method has become routine in the laboratory and provides a simple and effective means of producing high copy numbers of specific DNA sequences.

We start with one strand of the template DNA on which are the forward and reverse primer sites:

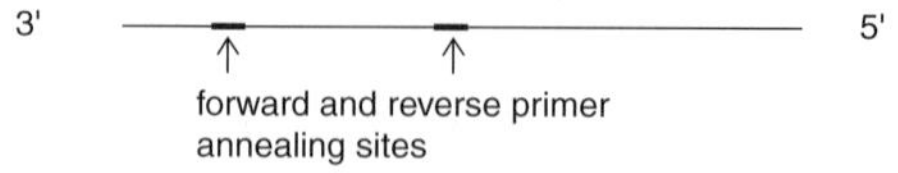

CYCLE 1

The forward primer anneals to one end of the target stretch and is elongated in the 5' to 3' direction by DNA polymerase:

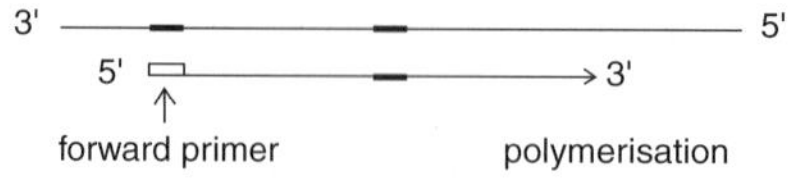

So, at the end of the first cycle there is the original template DNA plus what we will call an "overextended fragment":

One such overextended fragment is produced every cycle for each DNA strand, so that if we run PCR for, say, 35 cycles there will be 35 such overextended fragments per original template strand.

But what happens to overextended fragments in the next cycle?

CYCLE 2

The reverse primer anneals to the overextended fragment and is elongated by DNA polymerase - but only as far as the forward primer site, where the fragment ends. This produces our desired fragment, bounded by the forward and reverse primers.

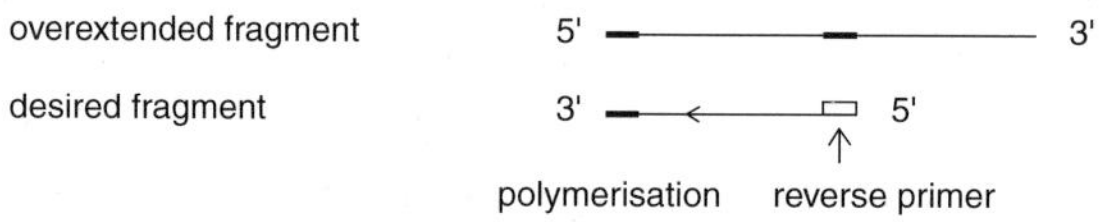

Each overextended fragment produced from the template DNA will produce perfect fragments in each remaining cycle. However, it is what happens to the perfect fragments in the third and subsequent cycles that really boosts the numbers.

CYCLES 3–35

In cycle 3, the perfect fragment makes a copy of itself; then in the next cycle both these copies copy themselves and in the following cycle all four of *those* copy themselves, and so on - this is the "chain reaction" element of PCR.

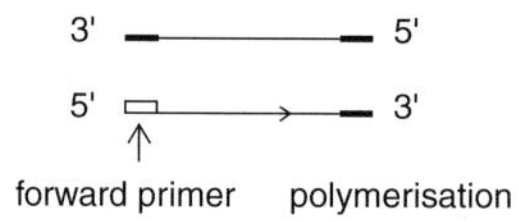

The first perfect fragment, produced in cycle 2, will double in each of the remaining 33 cycles, giving 2^{33} or 8,589,934,592 perfect fragments. But of course it's not just the first perfect fragment that doubles every cycle - so do all subsequent perfect fragments produced from each overextended fragment each PCR cycle. So, in theory, after the first two cycles the number of perfect fragments grows super-exponentially, to give a total of $(1 \times 2^{33}) + (2 \times 2^{32}) + (3 \times 2^{31})......+ (33 \times 2^{1}) + (34 \times 2^{0})$ = 34359738332 perfect fragments. It helps to understand these numbers if we realise that the total number of fragments doubles each cycle, so that after n cycles we have 2^n-pieces of DNA, of which one will be the original strand and n will be overextended fragments.

Thus, in theory, if we start with one DNA template strand, by the end of 35 cycles we have:

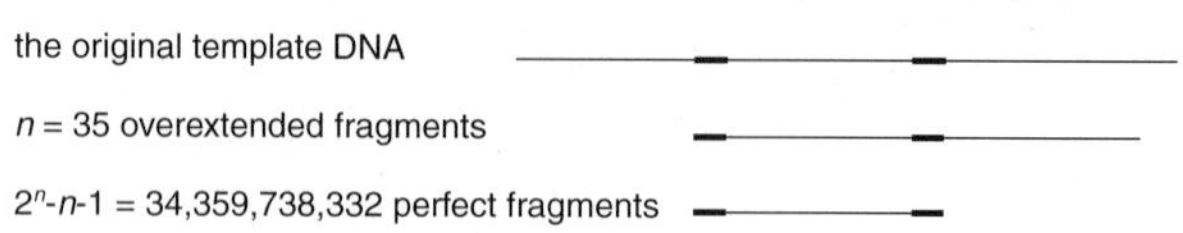

In practice these numbers are not achieved because dNTPs and primers run out and the DNA polymerase is not 100% efficient, but they serve to illustrate the DNA amplifying power of the Polymerase Chain Reaction.

Fig. B2.2b How the polymerase chain reaction (PCR) method works. To simplify the explanation the copying of only one of the strands of DNA is illustrated – the process is identical for the other strand. As the template DNA is normally double stranded each piece of template DNA would produce double the number of perfect fragments illustrated in this figure.

The sizes of pieces of DNA produced from cloning or PCR can be determined by subjecting the DNA to electrophoresis (Box 2.3) alongside known size-standards. Since electrophoresis separates fragments based on their sizes, it can be used to purify DNA fragments. For example, the results of a PCR reaction can be run (electrophoresed) on an agarose gel. The DNA is stained during or after electrophoresis with ethidium bromide which fluoresces under UV light. Hopefully there will be a nice bright band of the right size, our desired PCR product, which can then be cut out and the DNA extracted from the gel. We thus have the desired PCR product without leftover components of the PCR reaction, such as primers, which might have interfered with later DNA sequencing.

Box 2.3 Electrophoresis

Electrophoresis is used to separate molecules by size. It works on the principle that charged molecules, such as proteins or DNA, will be drawn through a slab of gel when a current is passed across it. A number of different gel types can be used, such as hydrolysed starch, cellulose acetate, agarose or polyacrylamide. Polyacrylamide gels are normally oriented vertically while other gel types are usually positioned horizontally. Polyacrylamide gel electrophoresis is sometimes known by the acronym PAGE. In vertical polyacrylamide gels, samples are placed at the top of the gel and are separated from one another by a comb-like structure, or by spacers. In horizontal gel systems, samples are inserted into slots in the gel close to, or at, one end. An example of a horizontal starch gel apparatus is given in Figure B2.3a.

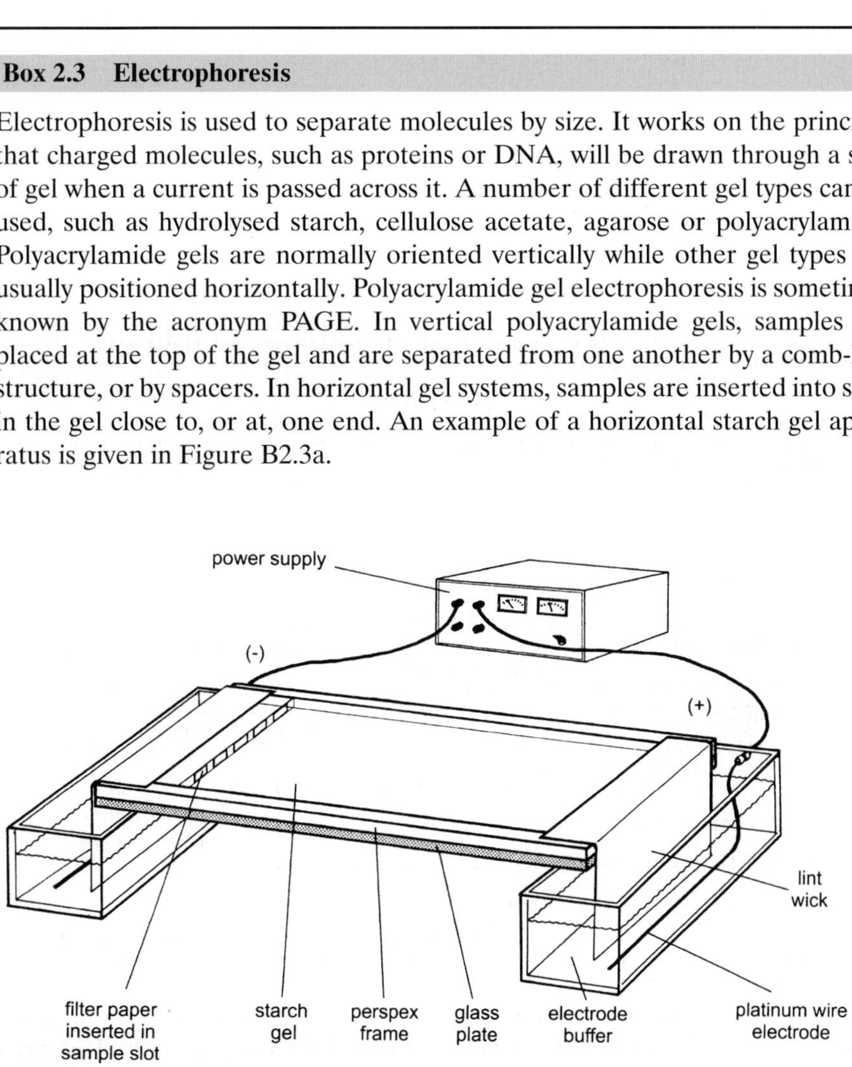

Fig. B2.3a Apparatus for the separation of allozymes by horizontal starch gel electrophoresis (courtesy Chris Beveridge).

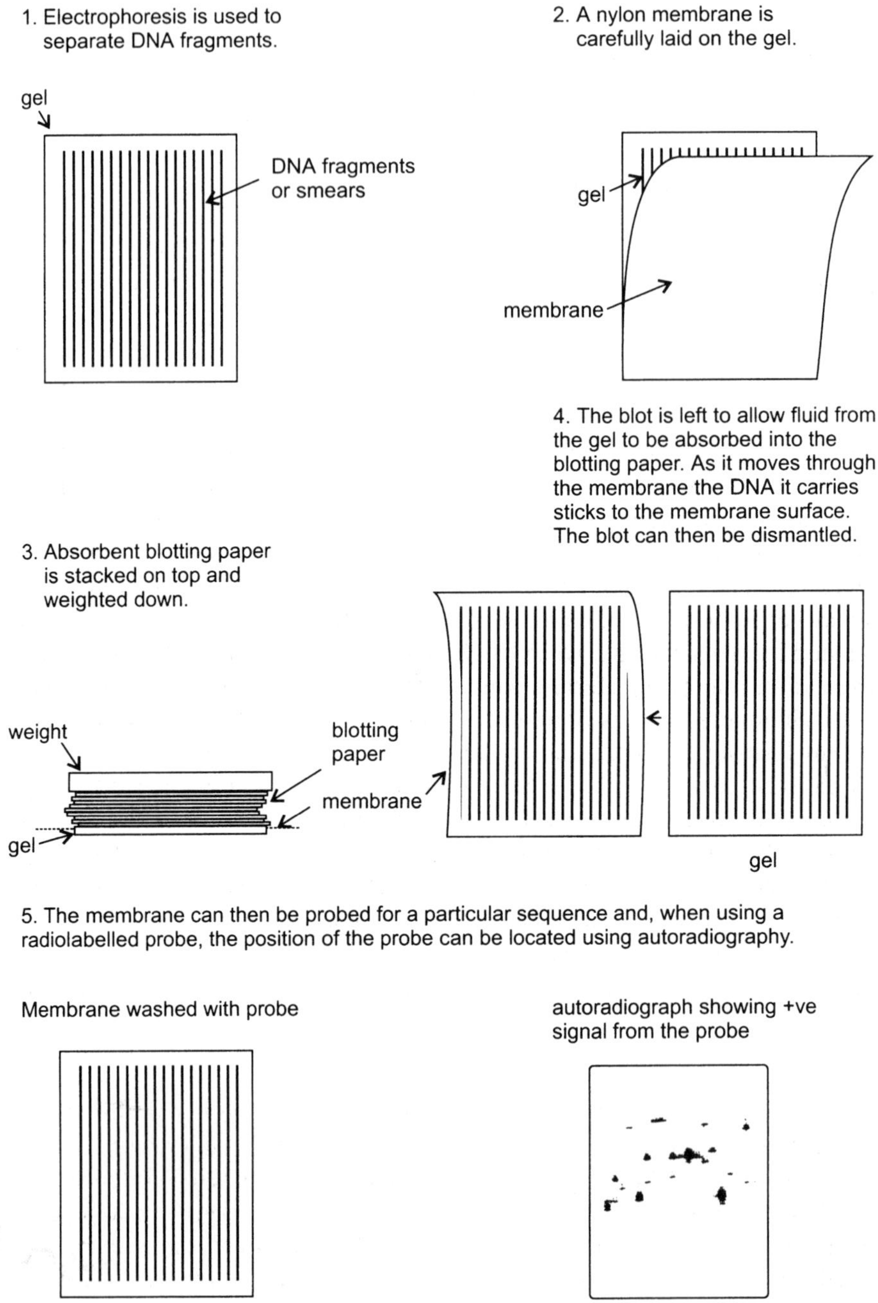

Fig. B2.3b The Southern blotting method of transferring DNA from a gel to a membrane.

The strength of gels can be adjusted to make the pore size similar to the size of the molecules being separated so that some sieving effect can take place in addition to the electrical charge dragging the molecules through the gel. Because passing an electrical current through water changes the pH, the solution used to make electrical connection with the gel is always buffered.

Once the current has been run for sufficient time to separate the faster-migrating from the slower-migrating molecules, electrophoresis is stopped and the gels are prepared for visualisation of the resulting bands of protein or DNA. For proteins a general non-specific protein stain can be used, though in the case of enzymes the positions of the different bands (allozymes) on the gel are identified using substrate-specific stains. High concentrations of DNA are usually stained with ethidium bromide, which fluoresces under UV light, but lower concentrations require the more sensitive silver staining method. For very small quantities of DNA the most sensitive staining method is to use radiolabelling. In radiolabelling, a radioactive isotope of an element, such as sulphur-35 (^{35}S) or potassium-32 (^{32}P), is incorporated into the DNA before electrophoresis, then the gel is dried and placed adjacent to a sheet of film (the autoradiograph negative) and the radioactive decay of the element exposes the negative at the point of the signal. Automated DNA analysis machines use chemo-luminescent stains that can be read by a laser. This removes the risks of working with radioisotopes and, by virtue of different-coloured stains, enables more DNA sequences to be obtained from a single gel. DNA can be radiolabelled after electrophoresis, but this requires it to be transferred from the fragile gel to a more robust membrane by a technique known as Southern blotting (Fig. B2.3b) before hybridisation with single-stranded DNA complementary to the sequence of interest.

Various standards can be run on gels alongside samples for comparison. Dyes and samples from individuals of known genotypes are run on protein and enzyme gels, while DNA bands of known sizes (in base pairs, bp, or kilobases, kb) are used as molecular size standards in DNA electrophoresis. The sizes of DNA fragments run on sequencing gels are found by comparison with a known sequence that is run alongside.

When we have lots of high-quality copies of the target DNA in a pure solution, we can use a standard sequencing method (Box 2.4) to identify the precise sequence of the bases (A, C, G and T) along the DNA. Comparison of the sequence between individuals, between populations, between species or between higher order systematic divisions, provides information about the relatedness between these categories. Of course, different classes of DNA are needed to address these different levels of relatedness. Although we can generally assume that the chances of a point mutation occurring are the same anywhere along the DNA molecules that make up the genome of a particular species, the important question is what the consequences might be of such a point mutation.

Box 2.4 DNA sequencing

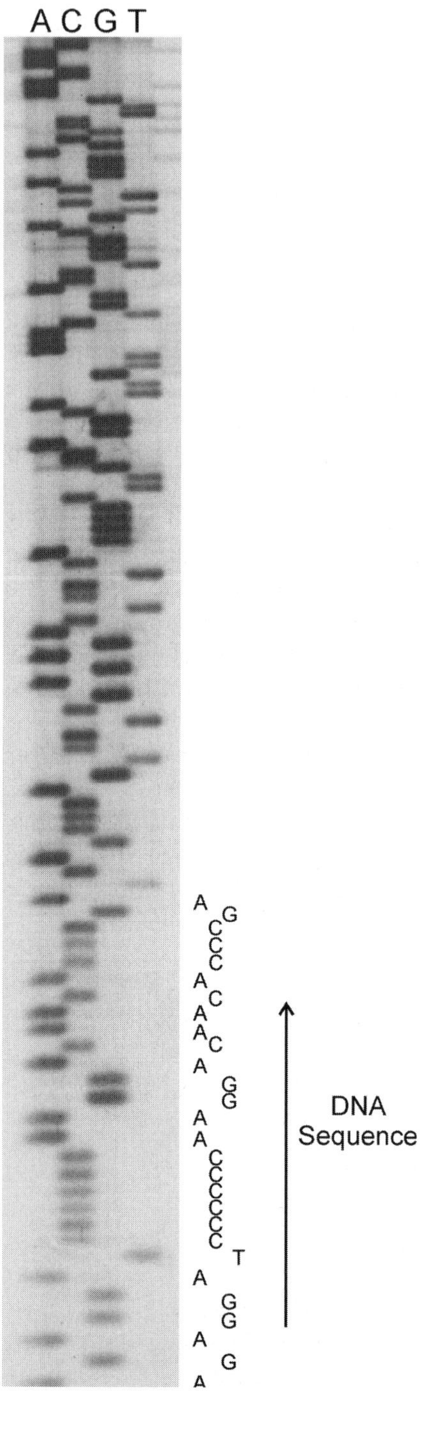

Fig. B2.4 Autoradiograph of a DNA sequence. Each base (A, C, G and T) is in a separate lane and the sequence is read from the bottom upwards.

Sequencing of DNA is now a highly automated and, in some cases, roboticised procedure. All well-provisioned large genetic laboratories will have their own sequencers and there are a number of commercial companies that provide a relatively cheap sequencing service to institutions such as marine stations or aquaculture institutions where genetic study is usually only a small part of their activities.

DNA sequencing uses DNA polymerase enzymes, such as those used in PCR, to copy the DNA strand but with two added twists. The first is that one of the dNTPs are fluorescently or radiolabelled, so that the copies can be visualised. The second trick is that we sabotage the copying process. We do this by introducing a small proportion of dideoxynucleotides (ddNTPs) along with the dNTPs. Like dNTPs, the polymerase joins ddNTPs to the new DNA strand, but unlike dNTPs they lack the bond which would enable another dNTP to be joined after them, so they stop the copying process. Sequencing one piece of DNA involves carrying out four separate reactions for Adenine, Cytosine, Guanine and Thymine using ddATP, ddCTP, ddGTP and ddTTP respectively. The proportion of ddNTP to dNTP is balanced so that copy strands are produced of many different lengths, from those that only extend a few bases from the sequencing primer to strands hundreds of bases long. But each will end in a ddNTP. So when we run the four reactions out on a sequencing gel, the ddATP reaction will produce bands of many lengths, but we will know that each band shows the length of a DNA fragment which ends with the nucleotide Adenine. Imagine that the sequence has Adenine occurring at the 2nd, 5th, 6th, 9th, 12th, 13th, etc. positions after a 20-base sequencing primer. In that case the A series will contain molecules of 22, 25, 26, 29, 32, 33, etc. bases long. Similarly, the ddCTP, ddGTP and ddTTP reactions will consist of molecules of lengths specific to the positions of the bases Cytosine, Guanine and Thymine, respectively, along the DNA. These four series of molecules are run in four lanes, side by side, down a high resolution polyacrylamide gel. The sequence of the DNA then can be read from these four ACGT lanes from the bottom of the gel upwards, as illustrated in Figure B2.4.

Sequence data have now been obtained from a great range of organisms and this information is collected together in DNA databases such as the one at EMBL in Europe and GenBank in the USA. Scientists have free access to these databases and powerful computer programs are available to analyse new sequences and to compare them with all other available sequences on the databases. This field of bioinformatics is rapidly expanding.

Let us first consider a mutation within the coded part (exon) of a gene that codes for an enzyme. We might expect such DNA mutations to have important effects. However, the mutation could occur at the third base of a codon and, because of the redundancy of the genetic code, will be unlikely to change the amino acid coded for. Alternatively, it could change one of the amino acids in the enzyme produced, but even this may not have any effect on the ability of the enzyme to carry out its cellular biochemical function. Nevertheless, *some* mutations within the exon of an enzyme gene are bound to have a deleterious effect such that individuals carrying that mutation produce an ineffective enzyme and are less likely to survive. Exceptionally, a mutation might be advantageous and improve performance of an enzyme. So enzyme exon DNA sequences are free to change slowly over evolutionary time, at a rate that is considerably less than the rate of mutation, and the rate varies between different enzymes depending partly on the specificity of their biochemical task in the cell.

What about DNA sequences which form part of an intron? These sequences are not translated into a protein product and so we would expect changes to have neither deleterious nor advantageous effects. Mutations at non-coding sites are effectively neutral and therefore are likely to accumulate without constraint over evolutionary time.

Finally, let us consider sequences that code not for proteins, but for the very RNA molecules which are involved in the process of translation of the DNA code. Here, almost every letter of the code is critical to the functioning of the RNA product and almost any mutation will render it non-functional. The strongly deleterious effect on any individual subjected to such a mutation means that the rate of evolutionary change of these parts of the DNA molecule is extremely slow. Such DNA is said to be highly conserved.

It follows from the three examples above that some regions of DNA are valuable for identifying evolutionary changes far back in time, while others will detect more recent changes.

DNA fragment size variation

At the beginning of this chapter we said that genetic variation can be measured and quantified at several levels. We have shown how we can determine the precise sequence of a length of DNA, and how it varies between individuals. Now we shall progress to see how differences between sizes of DNA fragments can be identified and used to address particular genetic questions. Techniques that fall into this category include those known by the acronyms RFLP, VNTR, DNA fingerprinting, RAPD and AFLP. Of these, VNTR markers (microsatellites in particular) have come to the fore in recent years as being the most generally useful, though the others all have their place in answering particular genetic questions.

Restriction fragment length polymorphisms (RFLPs)

We can make good use of fragments of DNA as genetic markers without going through the procedure of sequencing them. If we have a high copy number of a particular fragment produced by the cloning method (Box 2.1) or from the PCR machine (Box 2.2), this can be incubated with a number of different restriction endonucleases (REs) which will cleave it into a number of lengths depending on the position of the

Box 2.5 Restriction fragment length polymorphism (RFLP)

The fact that restriction enzymes will only cut DNA at specific sequences presents us with a simple way of identifying genetic variation caused by point mutations. Let's say we have a 2 kb length of DNA from an individual animal which can be amplified to a high copy number and we then incubate this amplified DNA in a microtube with a suite of restriction enzymes. The restriction enzymes will cut the DNA into a number of fragments that can then be easily size-separated on agarose gel and stained with ethidium bromide (Box 2.3). In other individuals of the same species, point mutations will have altered the sequence

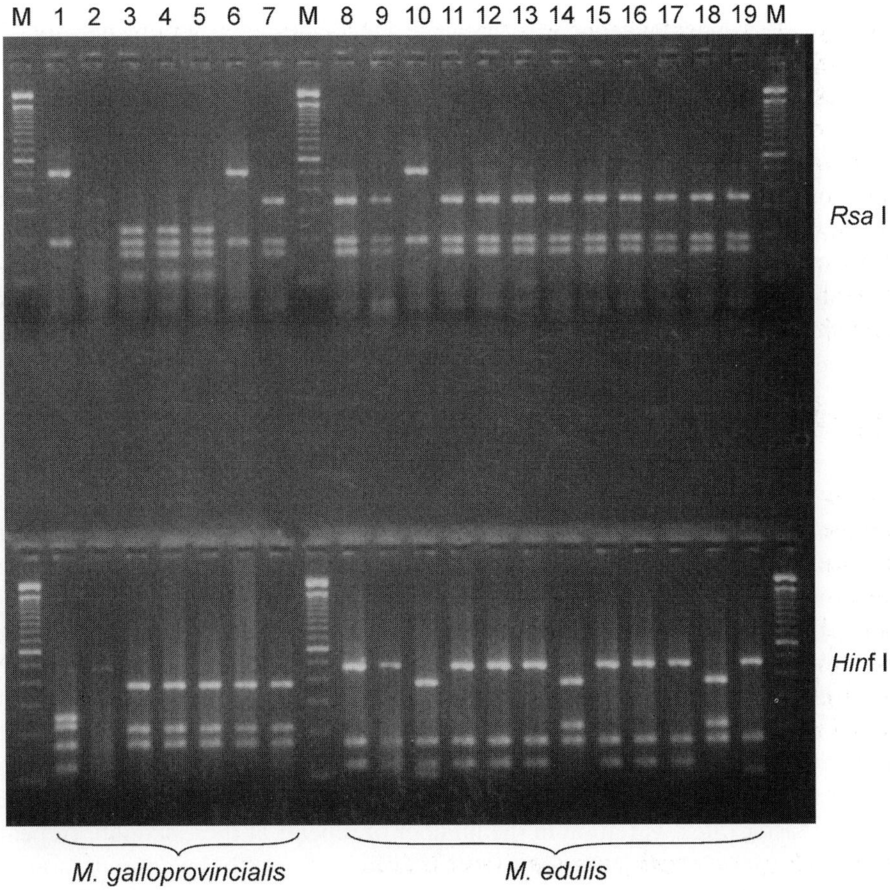

Fig. B2.5 Restriction fragment length polymorphism (RFLP) of a fragment of mitochondrial DNA from the mussels *Mytilus edulis* and *M. galloprovincialis*. The PCR product has been cut with the restriction endonucleases *Rsa*I (top) and *Hin*fI (bottom). Lanes 1–7 *M. galloprovincialis*, lanes 8–19 *M. edulis*. M = 100 bp ladder. Variation in the sizes of the fragments can be seen within species and between species. (Courtesy Dr Ann Wood.)

at one or more restriction enzyme cut sites, or may have produced a cut site where one was not present before. This will result in different individuals producing variation in the size and number of fragments when their DNA is incubated with this suite of restriction enzymes. Genetic variation identified in this way is called restriction fragment length polymorphism (RFLP) (Fig. B2.5).

RFLP data from a sample of a population can be analysed in two ways. First, all the different fragment patterns detected on the gels are counted and the frequencies of each determined. These RFLP frequency data can then be compared between populations. Secondly, RFLP data can be analysed on the basis of the proportion of nucleotides that differ between individuals. Of course, the number of nucleotides actually sampled is limited by the number of restriction enzymes used and the number of bases each enzyme has in its cut site. Nevertheless, such data are of value in establishing relationships between populations, species or higher taxa.

RE recognition sites. The various lengths produced can be separated by size and stained on an agarose gel (electrophoresis, Box 2.3). The same piece of DNA from different individuals will produce different sets of restricted fragments if there have been point mutations affecting the RE recognition sequences. In this way, polymorphisms can be identified based on the pattern of the size fragments on the agarose gel. More detail is given in Box 2.5. RFLP analysis is particularly useful for mitochondrial DNA (Box 2.6).

Variable number tandem repeats (VNTR)

Variation in the sequence of DNA can occur at certain sites by a method which is not point mutation. Spread throughout the genome are regions called variable number tandem repeats (VNTR), also known as simple tandem repeats (STR) or simple sequence length polymorphisms (SSLPs), which contain tandem (i.e. linked in chains) repeats of DNA sequences. The sequences may be very short (from 1 to 10 bp) or much longer, but the key feature of these tandem repeats is that the number of repeats can vary between individuals. It is thought that increases or decreases in the number of the repeats occur during copying by recombination or replication slippage and that these processes are not only independent of point mutations, but also occur at a much faster rates. Variation in the number of repeats at these satellite (repeated units 100 to 5000 bp), minisatellite (repeated units 5 to 100 bp) or microsatellite (repeated units 2 to 4 bp) loci can be very extensive in populations and provides a valuable tool for investigation of population genetic changes in the recent past. Microsatellite markers (Box 2.7) in particular are now used extensively for a number of reasons: because they are co-dominant (both alleles can be identified) and therefore can be analysed under the standard Hardy–Weinberg model (Chapter 3, Box 3.1); because, as 'junk DNA' they can usually be considered to be free of selec-

Box 2.6 Mitochondrial DNA extraction and analysis

It is possible to separate mtDNA from the nuclear DNA by differential centrifugation. A buffered chemical solution is used to break up (lyse) the cells. The resulting cell lysate is then centrifuged at a speed that is high enough to sediment heavier material such as the nucleus and larger cell debris. The supernatant, which contains the mitochondria and other cell organelles, is removed and centrifuged again at a higher speed to sediment the mitochondria. Further purification can be achieved by density gradient centrifugation where material is centrifuged through a series of layered density gradients. Once separated, the mtDNA can be extracted from the mitochondria using the standard phenol-chloroform extraction used for nuclear DNA.

Because mtDNA is a molecule of fixed length, and also because it is present in high copy number in cells, it is amenable to analysis without further amplification or preparation. Extracted mtDNA can be cut directly with restriction enzymes and the resulting fragments can be separated on an agarose gel and stained with ethidium bromide. Genetic variation between individuals is detected as sequence differences or as RFLPs (Box 2.5). The pattern of mtDNA restriction fragments from an individual is called its haplotype and the frequencies of particular haplotypes in a population are used to determine differences between populations. The degree to which mutational changes have separated different haplotypes – the nucleotide divergence – can also be quantified and used for population genetic or systematic purposes.

Box 2.7 Variable number tandem repeats (VNTR): microsatellites

The genomes of animals and plants contain regions that consist of a series of repeated units of DNA – VNTR. One type of VNTR – microsatellites – consist of dinucleotide (e.g. CACACACA), trinucleotide (e.g. GTAGTAGTAGTA) or tetranucleotide (e.g. TAGCTAGCTAGCTAGC) repeats. A microsatellite sequence identified in the DNA of the common cockle (*Cerastoderma edule*) is illustrated in Figure B2.7a.

The number of repeated units contained within a particular microsatellite locus can vary within a population, and this produces variation in the length of the locus. This variation can be detected by amplifying the locus using PCR, followed by electrophoresis (Box 2.3).

Isolation and identification of microsatellites in a species is done by first producing a library of recombinant clones (Box 2.1) containing fragments of DNA between 300 and 900 bp in length. DNA is extracted from an individual of the species, cut with REs and run out on an agarose gel against a size standard (Box 2.3). Fragments of a size between 300 and 900 bp are then extracted from

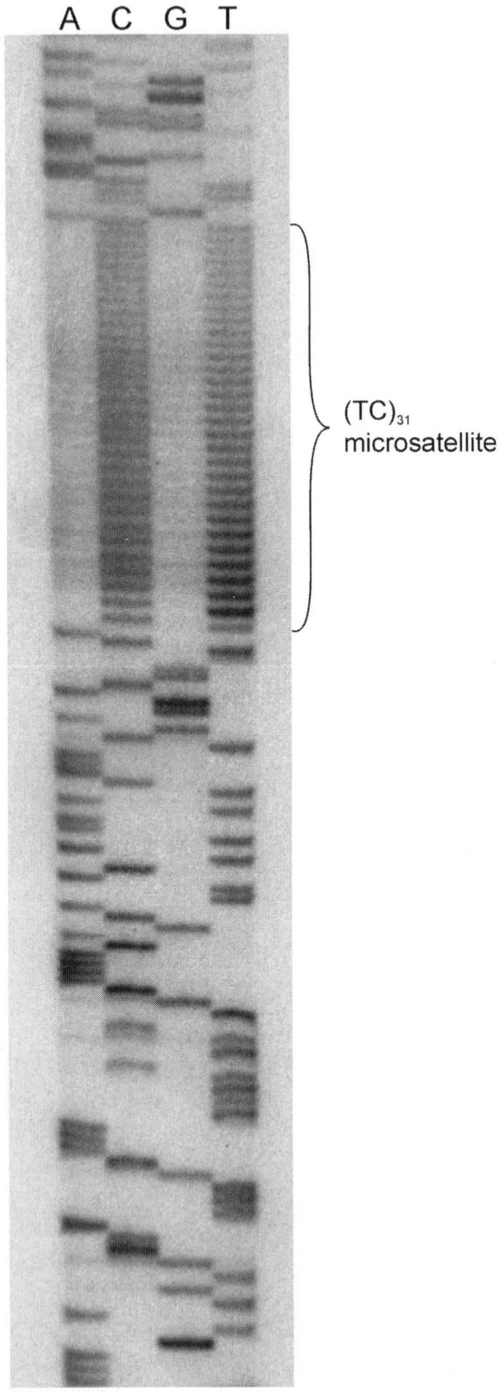

Fig. B2.7a A microsatellite sequence, $(TC)_{31}$, isolated from the European cockle, *Cerastoderma edule* (courtesy Dr Karen Abey).

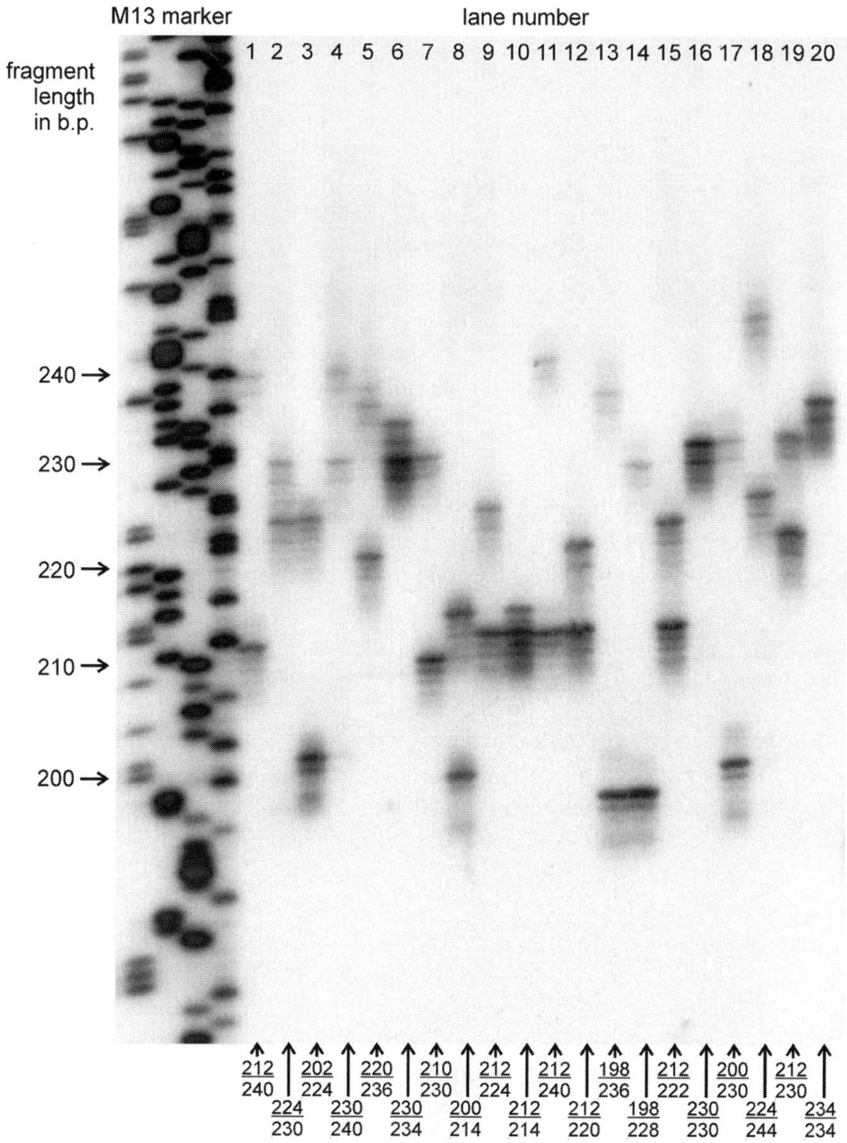

Fig. B2.7b Autoradiograph of microsatellite variation in the European oyster *Ostrea edulis*. Genotypes are scored as length in base pairs (bp) against the M13 size marker (courtesy Dr Halina Sobolewska).

the gel and these are used to make a clone library (Box 2.1). This library is then screened using complementary repeat probes, for example (GT)n to identify (CA) repeat microsatellites. Insert DNA from positive clones is sequenced to confirm the existence of a microsatellite within the fragment of DNA and to determine the flanking sequences. Primers are designed based on the flanking sequences and optimised for PCR (Box 2.2). The microsatellite locus is then PCR-amplified from template DNA extracted from individual organisms. The PCR products are run on a high quality polyacrylamide gel which enables the detection of fragments that differ by a single base pair in length. Scoring of genotypes is by radiolabelling, silver staining or laser detection in automated sequencers. However, there is often an effect – called 'stutter' – where slippage errors in replication during PCR produce fragments with one or two more, or one or two fewer repeats. This is most commonly observed with dinucleotide microsatellites and a little care is required in interpretation (Fig. B2.7b). Microsatellites are co-dominant, therefore both homozygous and heterozygous genotypes can be detected and microsatellite genotype and allele data can be analysed using the Hardy–Weinberg model (Chapter 3).

tive pressures; because of the high number of both loci and alleles at each locus; and, not least, because automatic DNA sequencers can be used for automated genotyping at microsatellite loci, vastly increasing the rate at which samples can be processed.

DNA fingerprinting

DNA fingerprinting can be thought of as a combination of RFLP and VNTR. First, genomic DNA is cut with a particular suite of restriction enzymes and differently sized fragments are separated by electrophoresis. The DNA is transferred from the fragile gel to a nylon membrane by the technique known as Southern blotting (Fig. B2.3b) and the membrane is then probed with a particular satellite repeat sequence which is common throughout the genome. Fragments that contain the repeat show up as a number of discrete bands after autoradiography. These banding patterns are so variable as to be in practice unique to each individual (the chances of a match between unrelated individuals are millions to one). Since the bands are inherited in a predictable fashion, DNA fingerprinting is a very accurate way of determining parentage and the forensic uses of this method are now well known.

Random amplified polymorphic DNA (RAPD)

The RAPD method (Box 2.8) is based on the principle that the shorter the length of the primers which are used in PCR, the greater is the chance that non-target sequences will be amplified. Using a single 10-mer oligonucleotide as the sole primer, PCR is conducted on raw DNA and the resulting fragments, which come from annealing of the primers all across the genome, are separated on agarose gel. Variations between individuals in the presence or absence of bands, reflect mutational difference at the primer sites. RAPDs suffer from the important criticism (among others) that they are not entirely reliable and repeatable.

Box 2.8 Random amplified polymorphic DNA (RAPD)

Primers designed for use with PCR (Box 2.2) are usually 20–25 bp in length to ensure that they will be specific to the particular DNA sequence being targeted. However, if a much shorter primer is used, say of 10 bp length, it is likely to anneal to many regions of the genome during the annealing step of PCR. If, by chance, the primer anneals to opposite strands of the DNA within a region up to about 3–4 kb in length, then a PCR product spanning the annealing sites will be produced. Some 10 bp RAPD primer sequences will produce no PCR product, while others may produce a number of different size fragments. The PCR products are run on agarose gel electrophoresis to identify any fragments according to their size.

How does this identify genetic polymorphisms? Consider the situation where a point mutation is present in the sequence at one of a pair of RAPD primer sites in some individuals in a population. PCR of the DNA from these individuals will not produce the fragment present in other individuals and therefore that band on the gel will be absent. Thus, presence or absence of particular bands can be scored as markers of genetic variation. This information of presence (+) or absence (–) of bands is called 'dominant' data. It is not possible to distinguish individuals that are homozygous (+/+) from those which are heterozygous (+/–) because both genotypes will produce a band on the gel. Allele frequencies, however, can be calculated based on the assumption that the locus is in agreement with the Hardy–Weinberg (H–W) model (Box 3.1, Chapter 3). The proportion of individuals scored for absence of the band (that is, –/– homozygotes) should equal the square of the (–) allele frequency.

When first developed, the RAPD technique seemed be a relatively quick and cheap method to obtain molecular genetic data that required no prior knowledge of DNA sequences. However, agreement with the H–W model has to be assumed and may not hold. Also, considerable care has to be taken to ensure that RAPD data are repeatable. Banding patterns are easily altered by subtle changes in the PCR conditions and some bands have been found which violate Mendelian inheritance. There is also the problem that it is difficult to identify bands that have been amplified from contaminating DNA (from parasites, gut contents, epibiota, etc.). For these reasons, the RAPD technique has fallen out of favour with many geneticists – though it is still widely used by those who work on clonal organisms as a way to identify particular clones.

Amplified fragment length polymorphism (AFLP)

A method which amplifies randomly selected fragments of the genome much more reliably than RAPD is the AFLP method (Box 2.9). AFLP is almost an inverse form of RFLP – the genomic DNA is cut into fragments with restriction enzymes and then

Box 2.9 Amplified fragment length polymorphism (AFLP)

Like RAPDs, the AFLP method provides size fragment-markers from DNA of unknown sequence. Unlike RAPDs, AFLP markers are extremely reliable and reproducible. DNA extracted from an organism is first cut using a pair of restriction enzymes which leave cohesive ends – one a frequent (four-base) cutter, the other a rare (six-base) cutter. The majority of fragments produced will have two four-base-cut ends, the minority will have two six-base-cut ends and the remainder will have one end with a six-base cut, and the other with a four-base cut. Two oligonucleotide 'adapters' designed to attach to the cohesive cut ends of both the six-base and four-base fragment terminals are then ligated to the resulting fragments. These adapters are used as the basis for polymerase chain reaction (PCR) primers that are also designed to overlap with the DNA fragment by one, two or three specified bases. The primer for the less frequent (six-base) cut adapter is radiolabelled before PCR amplification and polyacrylamide gel electrophoresis, so only those fragments which were cut by the six-base RE are visualised by autoradiography. By varying the number of specified bases on the primers the number of fragments amplified can be controlled. Typically 50–100 restriction fragments can be identified and separated in a single run of the AFLP method (Fig. B2.9).

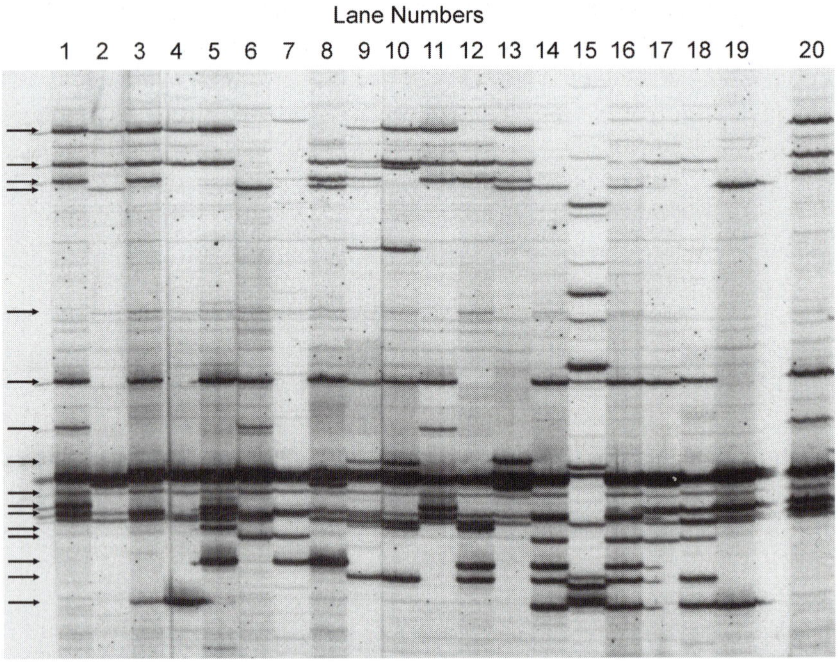

Fig. B2.9a Autoradiograph of amplified fragment length polymorphisms (AFLPs) in the Pacific oyster *Crassostrea gigas*. Arrows indicate bands which show variability between individuals (courtesy Dr Shelagh Malham and Dr Craig Wilding).

just a few of those fragments are selectively amplified using special radiolabelled PCR primers. The resulting fragments are then visualised by separation on a polyacrylamide gel followed by autoradiography. The products of AFLP – a series of bands of different sizes – are similar to the products of RAPDs and variation between individuals is based on the presence or absence of bands. Variation is the result of point mutation differences in the DNA sequence at the PCR primer sites.

Protein variation

So far we have considered genetic variation at the level of the DNA. However, DNA sequence variation, when transcribed, can give rise to differences in the resulting proteins. It is at this level that genetic variation begins to interact with the environment to affect the survivorship and reproduction of organisms and their genes.

Genetic variation at the level of proteins can be identified and quantified using electrophoresis (Box 2.3) to separate the different protein products of alleles followed by staining to visualise these protein products. It is possible to stain the gel to display all proteins, but it is more useful to take advantage of the substrate-specific catalytic abilities of the class of proteins known as enzymes. This involves using the specific substrate of an enzyme in a stain overlaid on the gel that will change colour where the substrate is altered by the enzyme. The position of any enzyme variants can therefore be located on the gel. These genetic variants of enzymes are known as allozymes and methods for detecting allozyme variation (Box 2.10) were first developed in the 1960s. Although nowadays regarded by some as an outdated method, the extensive allozyme data sets produced in the last 40 years of the twentieth century fundamentally shifted the ground upon which geneticists tread. From the practical point of view, allozymes enabled us to look at the genetics of natural populations of a whole range of aquatic organisms in a way that was never possible before. (The influence of allozyme data will become evident in several chapters of this book.) This is not to say that allozymes are of only historical interest – they are still a useful tool in answering many genetic questions, particularly given the equipment, time and money required to develop and use DNA techniques.

Although little used in fisheries or aquaculture research, the reader should be made aware of the technique of immunological testing which assesses the relationship between proteins on the basis of the relative strength of the antigen–antibody reaction that they will produce (Box 2.11).

Box 2.10 Allozymes

Allozymes are the products of genetic variation at enzyme-encoding loci. They should not be confused with isozymes that are alternative forms of an enzyme produced at different loci. In order to identify and score allozymes, a small piece of tissue is obtained from an organism and is ground up with a buffer solution in a microtube. This grinding releases the soluble proteins from the cells. After centrifugation, these proteins present in the supernatant are subjected to starch or cellulose acetate gel electrophoresis (Box 2.3) that separates the proteins on the basis of charge and size. This technique has been called the 'find 'em and grind 'em' method. Following staining for a particular enzyme, individual samples are classified according to the alleles present at the enzyme locus being investigated. Alleles are best identified by their mobility relative to a known standard – i.e. a band that moves 90% of the distance of the standard is called allele 90 – but they are sometimes just labelled numerically or alphabetically.

Interpretation of stained gels

Where an individual is a homozygote for an allozyme variant a single stained band will be seen on the gel, while heterozygotes, which contain two different variants, will exhibit two bands on the gel. Unfortunately this simple expectation is complicated by the quaternary structure of the protein. Figure B2.10a illustrates why heterozygotes for a dimeric enzyme exhibit three bands (two in the same positions as the homozygote bands + a hybrid band in between) and

Protein structure	Genotype	Protein subunits in the cells	Final proteins in the cell	Phenotype on gel
Monomeric	A	a	a	—
	B	b	b	—
	A	a	aa	—
Dimeric			ab	▬
	B	b	bb	—
	A	a	aaaa	—
			aaab	—
Tetrameric			aabb	▬
			abbb	—
	B	b	bbbb	—

Fig. B2.10a The effect of the final (quaternary) structure of proteins on the banding patterns of heterozygotes on electrophoretic gels.

why heterozygotes for a tetrameric enzyme produce five bands (two in the same position as homozygotes + three hybrid bands) on a gel after electrophoresis. Figure B2.10b shows the staining patterns from monomeric and dimeric loci.

Often, a particular enzyme will be coded for at more than one locus. This is common, for example, in salmonids where there has been a relatively recent evolutionary tetraploidisation event. In such cases, of course, the products of all the loci will stain on the gel. This can lead to difficulty in interpretation if the ranges of migration of the products of the loci overlap. The situation can get very tricky when there are duplicate loci for tetrameric enzymes whose products overlap. Note that additional hybrid bands will form between allelic products of both loci making a very complicated banding pattern produced in an individual that is heterozygous at both loci. Sometimes the problem can be alleviated by using particular tissues because some loci are tissue specific. For example, there are several lactate dehydrogenase (Ldh) loci in the trout, but the Ldh-5 locus is only expressed in eye tissue. Therefore, you can just pop out a trout's eye and grind it up to score for Ldh-5.

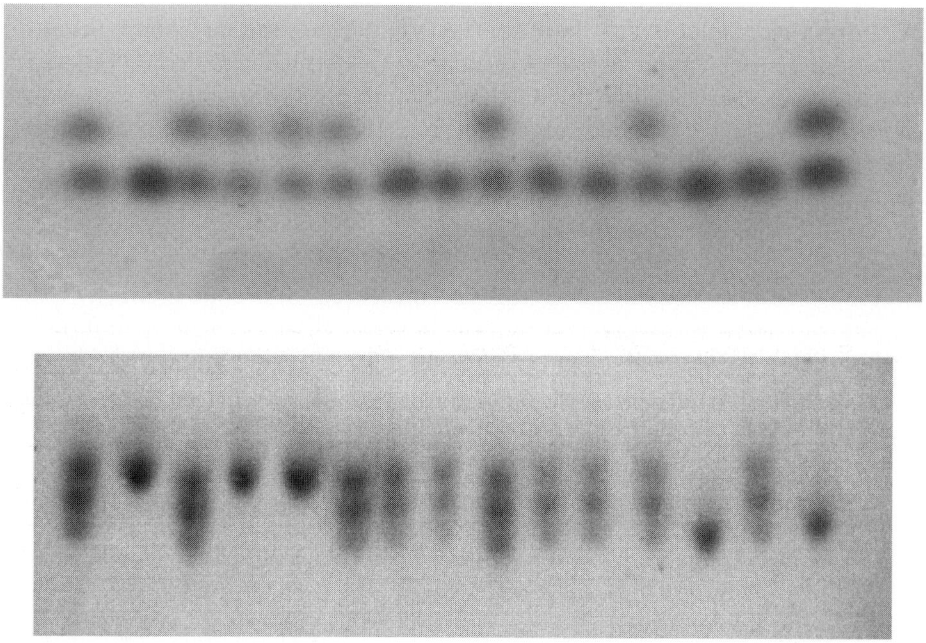

Fig. B2.10b Staining patterns exhibited by a monomeric enzyme, phosphoglucomutase (upper) and a dimeric enzyme, 6-phosphogluconate dehydrogenase (lower) following starch gel electrophoresis. Two alleles are present at each locus. Heterozygotes exhibit a two-banded pattern in monomers and a three-banded pattern in dimers.

Box 2.11 Immunological identification of proteins

Vertebrates have fairly-well understood systems in their blood for fighting and removing non-self or foreign proteins such as those in viruses and bacteria. Each protein has sites on its surface that are called antigenic determinants. When a non-self protein is detected the blood produces antibodies (immunoglobulin molecules) specific to the antigenic determinants of that protein. However, changes in the DNA coding for a protein can change the amino acid sequence in the protein and this, in turn, can change the nature of the antigenic determinants. This feature of proteins can be used for a number of purposes: to assess relationships between proteins extracted from different individuals or different taxa; to test for the presence of specific disease proteins in an animal; to identify foreign proteins coded for by transgenic DNA in genetically modified organisms (see Chapter 7).

It is rather complicated, but the method works as follows. Let us assume we are dealing with a comparison of a protein between two related species, A and B. First, a protein from an individual of species A is injected into the blood stream of a rabbit. The rabbit's blood will respond to this foreign protein by forming antibodies specific to the protein. If a sample of the rabbit's blood serum (blood cells removed), called antiserum, is then mixed in a suitable medium with the original protein from species A, a reaction will take place; the antibodies will link to the antigenic determinant sites on the protein to form aggregations which will precipitate from solution. If the antiserum is mixed with protein from species B, the amount of aggregation will depend on how many of the antigenic determinant sites are common between the two species. If they are very similar, there will be much aggregation; if they are very different, there will be very little aggregation. So the principle is that the amount of aggregation is proportional to the difference between the two proteins.

How then is the amount of aggregation measured? It is done indirectly using a method known as complement fixation. In serum there are a group of proteins called collectively 'complement' which bind on to antigen – antibody complexes and the more of these complexes there are, the more complement is bound up or 'fixed'. In the test, a known amount of complement is added to the antigen – antibody mixture and after a specified period of time, the amount of unfixed complement is assessed. Free, unfixed complement has the ability to lyse specially sensitised sheep red blood cells allowing the release of haemoglobin. Following this reaction, remaining blood cells are removed by centrifugation and the colour change produced by the released haemoglobin is measured in a spectrophotometer. So the colour change is inversely proportional to the amount of antigen–antibody aggregation; strong colour means there has been a weak reaction between the antigen and the antibody.

Table 2.1 provides a comparison of the values of various molecular techniques to address different problems in fisheries and aquaculture. The high value of VNTR methods in parentage and within species analysis is clear, but the extent to which allozyme data can be informative is also emphasised. Although RAPDs score quite highly for a number of approaches, there are real problems of repeatability and reliability with this method. AFLPs are now the recommended quick screening method for identifying quantitative trait loci (QTL, Chapter 5).

Phenotypic variation

There are very few examples of easily identifiable phenotypic variation in aquatic organisms controlled by single genes, or even pairs of genes. The best examples are found in the colouring of ornamental fish. It is interesting to consider just how lucky Gregor Mendel (the nineteenth-century discoverer of the method of genetic inheritance) was to have chosen to work with peas, which had a number of easily identifiable characters (round or wrinkled seeds; tall or short plants) each controlled by single genes. Such easily identifiable single-gene phenotypes are rare in most organisms. Because of the extensive development of DNA and allozyme technologies, the search for single-gene phenotypic variation is now uncommon. However, it is important to realise that the visual identification of varieties can be of critical importance

Table 2.1 Comparison of the value of genetic techniques in addressing different types of genetic problems in aquaculture and fisheries.

	DNA sequencing	RFLP	VNTR	RAPD	AFLP	Allozymes
Pedigree or parentage analysis	+	+	+++	+	+	++
Populations within species	+	+++	+++	++	++	+++
Genus and species level relationships	+++	++	+	++	++	+++
Quantitative trait loci	—	—	++	+++	+++	+
Cost	H	M	H	M	H	L/M
Tissue requirements	V	M	L	L	L	M
Codominant or dominant data	—	V	C	D	D	C
Neutral	V	V	Y	V	V	N

+++ = highly informative, ++ = informative, + = marginally informative, or constrained in some other way, — = not appropriate. H = high, M = moderate, L = low, V = variable, C = codominant data, D = dominant data, Y = yes, N = no.

to fish farmers without access to a modern genetic laboratory. Phenotypic rarities can provide high-value niche markets, and understanding how to get them to breed true requires genetic knowledge. Details of breeding schemes for phenotypic variations that are controlled at one or two loci are provided in Chapter 5.

Further reading

Avise, J.C. (1994) *Molecular Markers, Natural History and Evolution*. Chapman & Hall, London.

Brown, T.A. (1999) *Genomes*. Bios Scientific Publishers, Oxford.

Brown, T.A. (Ed.) (2000) *Essential Molecular Biology*, 2nd edn, Volumes 1 and 2, Oxford University Press, Oxford.

Carvalho, G.R. & Pitcher, T.J. (eds) (1995) *Molecular Genetics in Fisheries*. Chapman & Hall, London.

Majerus, M., Amos, W. & Hurst, G. (1996) *Evolution: the Four Billion Year War*. Longman, New York.

Turner, P.C., McClennan, A.G., Bates, A.D. & White, M.R.H. (1997) *Instant Notes in Molecular Biology*. Bios Scientific Publishers, Oxford.

Winter, P.C., Hickey, G.I. & Fletcher, H.L. (1998) *Instant Notes in Genetics*. Bios Scientific Publishers, Oxford.

Chapter 3
Genetic Structure in Natural Populations

What is a stock?

Worldwide, fisheries provide a source of employment for millions of people and a source of protein for millions more. The management of fisheries is therefore a critical activity. Unfortunately, the realisation that fish are a limited resource was slow to reach acceptance and there is not much evidence that attempts to manage fisheries have led to sustainability. In 1865, a group led by T. H. Huxley published the Sea Fisheries Commission Report into British fisheries, the first of its kind, in which it was recommended that there should be as few restrictions as possible on inshore and offshore fisheries and the fishing gear used. In effect, the conclusion of the report was that the supply of fish was inexhaustible. One hundred years later it was clear that this was patently not the case and in more recent times some of the world's major fisheries have collapsed through overfishing and poor management.

In order to manage a fishery effectively, a great deal of information is required. We need to know, for example, the size of a population, its age structure, the reproductive pattern of the species, the natural mortality rate, the rate at which fish are removed by fishing, and so on. Most of these factors can be, and are, determined from fishery statistics based on detailed surveys and analyses of landings. However, a critical requirement is to know whether the fish species exists as a single genetic unit or as a series of relatively genetically distinct groups. Is the species genetically homogeneous or is it genetically heterogeneous? If there are local genetically distinct groups, management strategies will need to be adapted to take this into account. The substructuring of fish into different, relatively genetically discrete groups may take various forms. For example, the migratory salmonids return to the rivers in which their parents spawned and this effectively creates genetic groups that are associated with particular rivers or even particular regions of a river. In another example, some lake fish have autumn and spring spawners and as long as this character is inherited, two genetically distinct groups can evolve. In many marine fish, there are particular sites where spawning aggregations occur and these can lead to genetic differentiation. In the case of sedentary shellfish, dispersal is usually via a larval phase and hydrographic factors will constrain the direction and distance of dispersal. Shellfish such as flat oysters with a short larval phase or direct development will be more likely to be genetically differentiated than shellfish such as mussels with an extended larval phase.

What should we call these genetically differentiated groups and what qualifies as 'genetically differentiated'? We shall see later how it is possible to quantify levels of genetic differences using various indices, but for the moment, let us consider the terminology that is available. Partly or strongly genetically differentiated groups within a species have been called 'varieties', 'races', 'types', 'strains', 'stocks', 'demes', 'populations' or 'subpopulations'. All of these terms have been used in the past to describe genetic differentiation within species but they have often not been clearly defined. In the case of fisheries, the word 'stock' has often been defined by fishery managers as a group of fish exploited by a specific method, or existing in a particular area. Although this may be convenient for the analysis of landings or catch and effort data, and for the enforcement of management measures such as quotas or net size, it may bear little relation to the true genetic substructuring of the fish species. The important point to realise is that there is a continuum of genetic differentiation. In some species (or in parts of their range) there may be clearly genetically distinct groups, while in others (or in other parts of their range) there will be groups which are virtually indistinguishable genetically. Thus, the use of a single word such as 'stock' to define the units of such substructuring is not often appropriate. Here, we will use the less prescriptive term of 'population' to define an intraspecific (within a species) group of randomly mating individuals which exists (and can therefore be sampled) in a defined geographic position or at a defined time. A key point to note here is that we are assuming that all individuals comprising a population are able to mate randomly with one another and therefore there can be no further genetic subdivision within a population.

In Chapter 2 the main types of genetic markers were outlined and we now need to show how these markers can be used to identify any population substructure within species of fish or shellfish. We will start with markers such as allozymes and microsatellites that are co-dominant, that is, they allow direct identification of homozygotes and heterozygotes. How do we handle this information on genotypes? First we should note that the process of meiosis breaks up existing genotypes and that new genotypes are reassembled when egg and sperm combine at syngamy. Therefore, although there will be the same range of genotypes in an offspring generation as in a parent generation, the two are not directly linked by inheritance. What is linked by inheritance, however, is the presence of an allele and the frequency of that allele in the population. And it is the allele frequency (or gene frequency) that is used to investigate the structure of populations within species.

How are allele frequencies estimated?

When genotypes at a co-dominant locus are scored from a sample of individuals, the frequencies of the alleles at the locus can easily be calculated from these data. Remember that every individual has pairs of chromosomes, so they must also have pairs of alleles at each locus, that a homozygote at a locus has two of the same allele, and a heterozygote has two different alleles. Although a diploid individual can only exhibit two alleles at a locus, there may be many different alleles at that locus present in the population as a whole. Allele frequencies are the critical starting point for all further analyses of genetics in populations. By convention, the frequencies of alleles at a locus are symbolised using the lower-case letters p, q, r, s and the frequency of an allele is given by:

$$p = \frac{2H_o + H_e}{2N}$$

where H_o = number of homozygotes for that allele, H_e = number of heterozygotes for the allele and N = number of individuals scored at the locus.

What is the relationship between alleles and genotypes?

Considering a single locus, it is convenient to envisage a population as a 'pool' of alleles. Haploid gametes are produced each generation and these gametes each contain a single allele from this pool. Let us assume that all adults in the population provide equal numbers of gametes and that each gamete has an equal chance of combining with all other gametes. Once we know the frequencies of the alleles in the population it is possible to calculate the frequencies with which the genotypes will be present in the offspring from that generation (Fig. 3.1). This principle was first quantified in 1908 by both Godfrey Hardy, a British mathematician, and Wilhelm Weinberg, a German doctor, and is called, unsurprisingly, the Hardy–Weinberg model (Box 3.1).

It can be seen from Figure 3.1 that with random mixing of eggs and sperm the proportions of the genotypes

	AA	AB	BB
will be:	p^2	$2pq$	q^2

which, in mathematical terms, is equal to $(p + q)^2$. As long as nothing upsets the Hardy–Weinberg model, allele frequencies should remain constant from generation to generation. The frequencies of the genotypes predicted by the model are dependent

Sperm \ Eggs	p	q
p	$p.p$	$p.q$
q	$p.q$	$q.q$

Fig. 3.1 Illustration of the principle behind the Hardy–Weinberg model. Here we consider the frequencies of genotypes produced by random combination of eggs and spermatozoa at a locus with two different alleles, A and B, at frequencies p and q.

Box 3.1 The Hardy–Weinberg model and causes of deviation from it

The Hardy–Weinberg model states that in a population of sexually reproducing diploid organisms, at a locus with two alleles A and B, at frequencies p and q, after one generation of random mating, the frequencies of the genotypes AA, AB and BB will be p^2, $2pq$ and q^2 respectively. The model is easily extended to more than two alleles such that the frequency of any homozygote equals the square of the allele frequency and the frequency of a heterozygote is two times the product of the two allele frequencies.

Statistical agreement with the model can be tested by the χ^2 (or G) goodness-of-fit test (for degrees of freedom see Box 3.5), but in genetic analysis computer programs an 'exact' test of probability is used. Agreement with the model is also indicated by the values of F_{IS} or F_{IT} (see Box 3.2, F-statistics). Significant deviation from the model can be caused by a number of factors:

1. **Selection.** Differential mortality of one particular genotype, or genotypes containing one particular allele, or one particular gamete type. Depending on which genotype(s) are the most fit, selection can be of the dominance, over-, under-, or semi-dominance types (Table B3.1).

Table B3.1 Types of selection. Maximum fitness of genotypes is 1, s and t are selection coefficients indicating that some proportion (s or t) of individuals die relative to the most fit genotype

	Fitness of genotypes		
	AA	AB	BB
Dominance	1	1	1-s
Over-dominance	1-t	1	1-s
Semi-dominance	1	1-s	1-(2s)
Under-dominance	1	1-s	1

2. **Lack of random mating.** This can take various forms:
   - *Inbreeding.* Mating between relatives, perhaps as a result of lack of dispersal, results in an excess of homozygotes (or deficiency of heterozygotes) relative to the model. The strongest form of inbreeding is self-fertilisation in functional hermaphrodites.
   - *Assortative mating.* Assortative mating occurs when individuals are more likely to mate with those that share some of their own characteristics – for example, a large individual may prefer to mate with another large individual. Assortative mating results in an excess of homozygotes for associated genes.
   - *Disassortative mating.* Disassortative mating is when individuals are more likely to mate with those that differ from them in some way – for instance, some organisms choose mating partners based on differences in their

immune characteristics to enhance the fitness of the resulting offspring. Disassortative mating produces an excess of heterozygotes compared to Hardy–Weinberg predictions.

3. **Migration.** Immigration into the population under study (between fertilisation and the time of sampling) of individuals from another population (separated from the first population geographically, or existing in the same location but spawning at a different time) with different allele frequencies can result in a significant deviation from the model. This is known as the Wahlund effect and it produces an excess of homozygotes. Similar results are produced where the sampled population actually consists of two different cryptic species with different allele frequencies at the locus under study.
4. **Null alleles.** Null alleles do not actually cause deviation from Hardy–Weinberg equilibrium, but they can make it appear that there is such a deviation. Where the product of a co-dominant allele cannot, for some reason, be visualised it is called a null allele. Failure to visualise an allele when scoring genotypes can lead to error because heterozygotes between a visualised allele and a null allele will be scored as homozygotes for the visualised allele. This results in an apparent excess of homozygotes against the model. Null alleles can be caused in allozymes by failure of the protein product of an allele to act as an enzyme. In microsatellites, null alleles can be caused by a mutation in one of the primer sites causing failure of PCR to make a product.

on the assumptions of equal and random parental contributions and equal and random mixing of gametes. However, if this is not the case, the model will not hold true. Also, we do not usually sample animals at the beginning of their lives; normally fish or shellfish are sampled as adults. In order for the Hardy–Weinberg model to be appropriate in populations of adults, further assumptions need to be made. We must assume that there is no selection that causes one or other genotype to suffer differential mortality. We must also assume that our sample taken does not include individuals that have migrated into the population from some other population with different genotype frequencies. So migration and selection are both factors that can cause deviation from the genotype frequencies predicted by the Hardy–Weinberg model. A further consideration is that very few fish or shellfish species exist as discrete generations, most populations being mixtures of animals from different overlapping generations. More details about deviations from the Hardy–Weinberg model are provided in Box 3.1.

How do allele frequencies change over time?

If a population is in agreement with the Hardy–Weinberg model then allele frequencies at a locus are not being driven (by selection or migration) to change significantly from generation to generation. However, there is always bound to be some variation in allele frequency from one generation to the next because of the element of chance. Contrary to the idealised situation required to make mathematical predictions, in real life all individuals of all genotypes will not produce exactly the same number of gametes. Neither are all gametes likely to undergo truly random mixing. Therefore there will be variation from the model in every generation. This is a fact of biological variability, called random genetic drift. In order to quantify this natural variation, allele frequencies can be given a variance as indicated in the formula:

$$\text{variance of frequency of allele} = \frac{p(1-p)}{2N_e}$$

where p is the frequency of the allele and N_e is the effective population size. The concept of the effective population size is an important one. It is the number of individuals in the population that contribute genetically to the next generation. It will exclude juveniles, individuals too old to reproduce and those that provide non-contributory gametes. These may be individuals that have infertile gametes, or those whose gametes never come into contact with the gametes of other members of the population due to geographic position or timing of spawning. Thus it can be seen how the effective population size could, in certain circumstances, be very much smaller than the total number of individuals in the population.

If the effective population size (N_e) in the expression above is a very large number, then the allele frequency variance will be very small. Conversely, allele frequency variance will be large when N_e is small. This means that natural fluctuations in allele frequency between generations will be much greater in small populations. We shall encounter this again when dealing with various aspects of population genetics in aquaculture.

Allele frequencies in a population can therefore change over time owing to random genetic drift. However, they may also change owing to pressure of selection (one genotype or one allele survives better than others at a locus) or to patterns of migration or dispersal that may fluctuate in direction or in strength over time.

How does population structure arise?

The differentiation of a species into genetically different populations is a fundamental part of the process of evolution. For a number of physical or biological reasons the distribution of a species may become fragmented. For example, during the last ice ages, aquatic species in the temperate regions of the northern hemisphere were driven south and many became restricted to local areas in southern habitats. Where species were fragmented into different areas, many of these populations remained isolated for long periods of time in such refugia. Following the retreat of the last glaciation some 10 000 to 13 000 years ago, aquatic and terrestrial species went through the process of gradual recolonisation of habitat and this has provided plenty of opportunity for further fragmentation. Once fragmentation has occurred, allele frequencies at most loci will be subject to random genetic drift in the fragmented populations. As we have seen, random genetic drift produces more-rapid allele frequency change in small, compared with large, populations. If only a few individuals are involved initially in founding a population then allele frequencies in the new population may be very different from the source population, and if numbers remain small further allele frequency changes may be rapid.

In addition to random genetic drift, localised adaptation will occur involving selection at some loci for particular characteristics and these will cause further differences between populations. These processes of random genetic drift and adaptation will tend to increase differences between populations while migration or larval dispersal between populations will tend to reduce it. So current population differentiation in a species is essentially the result of the historical interplay between the environment and the forces of dispersal, adaptation and random genetic drift.

How are genetic markers used to define population structure?

Some of the genetic markers described in Chapter 2 are ideal for identifying and quantifying the level of population structure within a species. Most data exist for allozymes, but there are plenty of examples of the use of mtDNA variation to look at stock structure, while microsatellites are emerging as the markers of choice in current studies.

In any population study, the ideal first step would be to collect samples of the species across its entire range to estimate genetic differentiation within the species as a whole. In fact, this is seldom done. For reasons of economy most genetic studies related to fisheries have tended to focus on limited sampling in specific areas where there has been a commercial interest.

Depending on the type of marker employed, genotypes (allozymes, microsatellites) or haplotypes (mtDNA) are scored for the individuals sampled and the data are analysed in a variety of ways to quantify levels of genetic variation between populations. One commonly-applied type of analysis, called *F*-statistics, was developed by Sewall Wright in the first half of the twentieth century (Box 3.2). Because the development of mathematical models requires certain concepts or parameters to be fixed at the start, various assumptions had to be made in the development of *F*-statistics and it is important to consider what these were.

First, Wright made the assumption that all populations were of the same size, that is, they consisted of approximately the same number of reproductively active individuals. So, in this instance, population size is not the same as geographical size. How valid is this assumption? Well, it is easy to imagine situations, particularly in a heavily-fished species, where population size could be very different between populations. Also, in sedentary shellfish, the size of populations will be constrained by, among other factors, the availability of suitable habitat.

The second assumption was that there was an equal possibility for any population to exchange individuals with any other population. Anyone who has some knowledge of aquatic habitats such as lakes, rivers and oceans will immediately see the difficulty with this assumption, as only in very special circumstances is it likely to hold true.

Box 3.2 *F*-statistics

Sewall Wright introduced *F*-statistics as a method of describing the population genetic structure of diploid organisms. There are three indices, F_{IS}, F_{IT} and F_{ST}. '*F*' comes from 'Fixation', while 'I' refers to the individual, 'S' to subpopulation and 'T' to the total population. Hence F_{IS} refers to the Hardy–Weinberg distribution (or otherwise) of genotypes of individuals within subpopulations, F_{IT} to the distribution of individual genotypes within the total population and F_{ST} to the genetic differentiation of subpopulations within the total population.

F_{IS} is defined as 'the correlation between homologous alleles within individuals with reference to the *local* population' and F_{IT} as 'the correlation between homologous alleles within individuals with reference to the *total* population'. Effectively these describe whether the proportion of homozygotes (homologous alleles within individuals) in a sample is in agreement with the proportion expected under the Hardy–Weinberg model. For F_{IS} the samples tested come from local populations and the overall value for F_{IS} is a simple arithmetic mean of the values for each local population. On the other hand, F_{IT} is calculated from the total population, that is all samples from the local populations pooled together and treated as a single sample.

F_{IS} and F_{IT} are sometimes known as fixation indices (F_I) and are calculated from the number of heterozygotes observed (H_{obs}) and the number of heterozygotes expected under the Hardy–Weinberg model (H_{exp}):

$$F_I = 1 - (H_{obs} / H_{exp})$$

This formula also describes F_{ST} if we use the heterozygosities of the subpopulation and total population as the observed and expected values. Positive values of fixation indices demonstrate an excess of homozygotes (positive correlation between homologous alleles) or, conversely, a deficiency of heterozygotes, relative to the Hardy–Weinberg model. Where F_{IS} is positive in a local population this could be due to inbreeding (or other causes, see Box 3.1) and this index is often labelled an inbreeding coefficient. Where F_{IT} is positive, this could be the result (among other things) of differences in allele frequencies between local populations (the Wahlund effect, see Box 3.1).

F_{ST} is an index of genetic differentiation that describes how much variation in allele frequencies is present between the local populations. The three indices, F_{IS}, F_{IT} and F_{ST} are interrelated according to the formula:

$$1 - F_{IT} = (1 - F_{ST})\,(1 - F_{IS})$$

and F_{ST} can also be interpreted as the variance (V_p) of the allele frequencies among local populations relative to the maximum value possible based on the mean allele frequency (P) across all local populations:

$$F_{ST} = V_p / (P\,(1 - P))$$

Masatoshi Nei developed a similar statistic to F_{ST} that he called the coefficient of genetic differentiation, G_{ST}. It is calculated on the basis of heterozygosities where H_S is the average heterozygosity across local populations, and H_T is the heterozygosity of the total population:

$$G_{ST} = (H_T - H_S) / H_T$$

These statistics F_{ST} and G_{ST} are valuable measures of population subdivision and can also be used to estimate the amount of gene flow between the subdivided populations in the form of $N_e m$, the number of migrants between the subdivided populations each generation, since:

$$F_{ST} (= G_{ST}) = 1 / (4N_e m + 1)$$

therefore

$$N_e m = (1 - F_{ST})/4F_{ST} \quad \text{and} \quad m = (1 - F_{ST})/(4F_{ST}N_e)$$

where N_e is the effective population size and m is the proportion of migrants arriving into each of the subdivided populations in each generation.

In most cases, movement of larval individuals between populations is likely to be unidirectional rather than multidirectional due to the effects of currents and there are very few fish species where exchanges of juveniles or adult individuals between populations would be expected to be random in both directions. These two theoretical assumptions are a required part of the 'island model' which Wright used as a basis for F-statistics. It was further assumed that changes in allele frequency at most loci over time were essentially due to random genetic drift rather than to selection.

Wright's F-statistics (Box 3.2) provide answers to two different questions. The first question is: for the loci scored, are the genotypes in the proportions predicted by the Hardy–Weinberg model? F_{IS} provides a measure of this agreement for a single population and F_{IT} for all the populations combined. F_{IS} and F_{IT} can vary from –1.0 through zero to 1.0 and exact agreement to the Hardy–Weinberg model equals zero. The second question is: for the loci scored, are the allele frequencies different between various populations? F_{ST} provides a measure of this population differentiation and ranges from zero, where all populations have the same allele frequencies at all loci, to 1.0 where all populations are fixed for different alleles at all loci.

The Japanese geneticist Masatoshi Nei developed G_{ST}, the coefficient of genetic diversity, which is an equivalent index to Wright's F_{ST}. It is calculated slightly differently (Box 3.2), but is essentially addressing the same question. Because these indices represent the genetic differentiation between populations, it is possible to suggest how many individuals might be being exchanged per generation between these populations in order to produce the amount of differentiation observed. Using the formula $m = (1 - F_{ST})/(4F_{ST}\,N_e)$ given in Box 3.2 and taking the example of a large effective population size of 1 million ($N_e = 10^6$) with a value of F_{ST} of 0.000025, we would calculate that about 1% of individuals in the population were migrants

(m = 0.01). Thus N_em, the number of migrants per generation, would be 10,000. It is easy to see that with that high number of migrants per generation there is very little possibility of genetic differentiation developing between populations. On the other hand, with a smaller effective population size ($N_e = 10^3$) and an F_{ST} of 0.2, we would estimate a very low proportion of migrants (m = 0.001), with an N_em of 1.0 per generation.

With the exchange of individuals comes the exchange of genes and the greater the exchange, the less will be the genetic differentiation between the populations. Although there is no hard and fast rule, it is accepted that important or significant genetic differentiation is only likely to arise between populations when fewer than one individual per generation, on average, is being exchanged. Again, this is based on the assumption that changes in allele frequencies are based on random genetic drift alone. Obviously, at some loci, there could be significant changes due to selection and such changes would give the false impression that there is less migration between populations than is actually the case.

There are alternative ways of estimating genetic differentiation besides using Wright's F_{ST} or Nei's G_{ST}. It is possible simply to test for the heterogeneity of allele frequencies at each locus across all populations using Contingency Table tests. Traditional testing would involve the use of the χ^2 (chi-squared), or G, test, but nowadays, with the enormous computing power available in desktop PCs, much safer 'exact' tests can be employed and these are used in modern genetic analysis computer packages. What this test tells us is whether there is significant heterogeneity in allele frequencies across all populations. Selective removal of populations that look particularly different, followed by retesting the data, can reveal further detail. Care is needed to avoid the type I statistical errors associated with the use of several tests of the same hypothesis (Box 3.5, Statistical problems).

Finally, allele frequencies can be used to calculate pair-wise genetic differences between populations, or species, using formulae developed by a number of scientists. Nei's genetic identity (I) and genetic distance (D) (Box 3.3) are the most commonly used. Once again, a major difficulty faced by geneticists in trying to describe what can be an extremely complex situation by a single statistic is the problem of the neutrality of the gene loci used. On the one hand, genetic variation at many loci could be adaptive such that selection operates and particular alleles or genotypes are favoured in particular situations. Alternatively, allelic variation at most gene loci could be neutral, that is, it varies by random genetic drift and is not subject to selection. For most genetic indices, the usual assumption is that all of the alleles at all of the loci included in the determination of the index are neutral.

Levels of genetic differentiation in aquatic organisms

Many studies have been carried out to estimate the amount of genetic subdivision within aquatic species. Table 3.1 illustrates levels of population substructure in various groups. On average, these levels of G_{ST} (which is equivalent to F_{ST}) are not too far from 0.2, suggesting that, in general, subdivision of populations within species is

Box 3.3 Genetic distance measures based on allele frequencies

Nei's coefficients of genetic identity (I) and genetic distance (D) are indices of the genetic similarity, or distance, between two groups of organisms. These two taxa can be different populations of the same species, different species, or different genera. For a single locus, the coefficient of genetic identity (I) between two taxa is given by:

$$I = \frac{\Sigma x_i y_i}{\sqrt{(\Sigma x_i^2 \ \Sigma y_i^2)}}$$

where x_i and y_i are the frequencies of the ith allele in populations X and Y respectively. This method is based simply on allele frequency differences between taxa and can be applied to co-dominant data from allozymes and microsatellites and also to dominant data from RAPDs or AFLPs.

For multiple loci, the overall identity is:

$$I = \frac{J_{xy}}{\sqrt{(J_x J_y)}}$$

where J_{xy}, J_x and J_y are the arithmetic means across loci of $\Sigma x_i y_i$, Σx_i^2 and Σy_i^2 respectively. I ranges from one, which indicates that the taxa share the same alleles at the same frequencies, to zero, where no alleles are shared between the taxa.

Genetic distance (D) is simply the negative natural log of I:

$$D = -\ln I$$

When I is 1, D is 0 and there is no genetic distance between the taxa. Increasing values of D indicate increasing separation between the taxa and, because D can be considered to represent the mean number of codon substitutions per locus, it is possible to suggest a time since divergence of the two taxa. Where T is the time in years since divergence, Nei proposes that:

$$T = 5 \times 10^6 D$$

but other authorities disagree, both on the figure of five million years per unit of Nei's D and also about whether we should make the assumption that there is a constant and regularly ticking 'molecular clock' over evolutionary timescales.

Where measures of genetic similarity have been calculated for all possible pairs of populations the estimated relatedness between the various groups can then be illustrated using a dendrogram. There are a number of methods, the commonest being unweighted pair-group method of averages (UPGMA) cluster analysis. The dendrogram consists of a horizontal scale of genetic distance or similarity with vertical lines linking different populations or operational taxonomic units (OTU) at points on the scale at which they diverge genetically (see, for example, Fig. 3.3).

Table 3.1 Mean levels of population substructure in aquatic groups based on values of Nei's coefficient of genetic differentiation, G_{ST}, estimated from allozyme data

Taxonomic group	G_{ST} (± standard error)	Number of species
All vertebrates	0.202 ± 0.015	207
Fish	0.135 ± 0.040	79
All invertebrates	0.171 ± 0.020	114
Crustaceans	0.169 ± 0.016	19
Molluscs	0.263 ± 0.036	44

close to that which allows genetic drift to change allele frequencies in local populations in the face of gene flow by migration. However, these average values hide a wide range of variation in population structure between species. For example, let us take two contrasting cases – the salmon, *Salmo salar*, and the mussel, *Mytilus edulis*. The salmon has a life history involving extensive juvenile and adult migration, with mature adults returning to their natal rivers to spawn. Thus, although salmon can be found over a very wide area, they are genetically very much restricted to local populations in their own rivers as this is where the transmission of genes between generations occurs. There are strong barriers to gene flow because very few fish return by mistake to non-natal rivers. Each local population consists of relatively few individuals and the population can be sometimes even further subdivided into year classes that return to spawn in the same year as one another. Values of F_{ST} for salmon, calculated from extensive allozyme data, are in the region of 0.4 and indicate that, as expected, there is very little gene flow ($N_e m = 0.38$) and strong genetic differentiation into local populations in this species. On the other hand, the mussel is a sedentary organism and adults never actively move more than a very short distance. As with salmon, mussels release eggs and spermatozoa into the water, but, in contrast to salmon, mussel larvae are small and planktonic and are at the mercy of currents that can disperse them over great distances. Larval life, from egg to metamorphosis, in mussels lasts around 4 weeks and can be further extended in two ways. First, if a suitable habitat for settlement is not discovered, metamorphosis can be delayed for more than a week, allowing further dispersal. Secondly, once metamorphosed, these very young mussels (spat) can detach, secrete a very thin byssal thread and be transported in currents by a process known as byssal drifting, which is analogous to the aerial dispersal of spiders by gossamer threads. Typically F_{ST} values from allozyme studies in mussels are very low (<0.001) and clearly reflect extensive gene flow ($N_e m > 250$) and a lack of any population substructure.

It is also interesting to compare genetic differentiation in marine fish such as Atlantic herring (*Clupeus harengus*) with that in *anadromous* fish such as salmon. The herring has a very large effective population size because all fish return to the same region to spawn and there are therefore few, if any, barriers to gene flow. Average F_{ST} for herring is 0.01 and $N_e m$ is 24.8.

Therefore, the conclusion is that where the effective population size and/or migration is large, gene flow will tend to dominate over random genetic drift, there will be little differentiation and F_{ST} will be close to zero. Where $N_e m$ is small, random genetic drift will tend to dominate over gene flow, allele frequencies will differ strongly between populations and F_{ST} will be large.

Although information about population differentiation is very valuable to fishery managers, it is important to recognise that a *lack of evidence* for local genetic populations does not always mean a *lack of substructure*. If populations are genuinely genetically isolated from one another then we would expect random genetic drift to be acting to change allele frequencies at all polymorphic loci. However, for reasons of chance, not all loci will show different allele frequencies in different populations. If the loci screened in limited surveys happen to be these invariant loci then the true variation remains hidden. It is therefore important to examine sufficient loci to be sure that an apparent lack of genetic differentiation can be relied upon.

Allozyme studies suffer from another cause of failure to detect genetic isolation because of their very nature. As enzymes, they are vulnerable to selection operating at the biochemical level, either directly on them or on the biochemical pathways within which they operate. If the selection is such that it is in the same direction in two genetically isolated populations, that is, an allele is at high frequency because the individuals carrying it are favoured in both populations, then the allozyme data will be interpreted as indicating that the populations are not genetically different. On the other hand, two populations with habitat differences but which regularly exchange genes via larval flow may have different allele frequencies at a locus which is under selection where one allele is favoured in one population but is deleterious in the other. Differential mortalities in the two populations will cause allele frequency differences between samples taken from the two populations and this will give a false impression about the degree of reproductive separation of the two.

As an example, we show allele frequency data at a protein locus (*PT-A*) for populations of the queen scallop (*Aequipecten (Chlamys) opercularis*) around the British Isles and northern France (Fig. 3.2). There are three alleles at the locus, differing in frequency in different regions. In the south and west, populations are characterised by a high frequency of one allele, another predominates in the north, while the third is the most frequent allele found in the Irish Sea, on the east coast of Britain and in the Bay of St Brieuc. The differences in allele frequencies between the main regions are clearly quite large and the immediate conclusion might be that there are several different populations or 'stocks' of the queen scallop around the British Isles and northern France. However, this begs the question: are these differences really the result of lack of gene flow between populations, allowing the observed genetic differences to have accumulated through genetic drift? Or could there be plenty of larval exchange between populations and the differences be simply the result of selective mortality of certain of the *PT-A* genotypes between the larval and adult stage?

Knowing the life history of these scallops we can say that larvae certainly live long enough to allow exchange between these populations, but of course whether this actually happens depends upon the direction and speed of ocean currents. In the case of the Bay of St Brieuc, there is strong evidence of a cyclical gyre that will tend to retain scallop larvae within the confines of the Bay. Also, in the Irish Sea there are a number of important fronts affecting the flow of water which build up at certain times

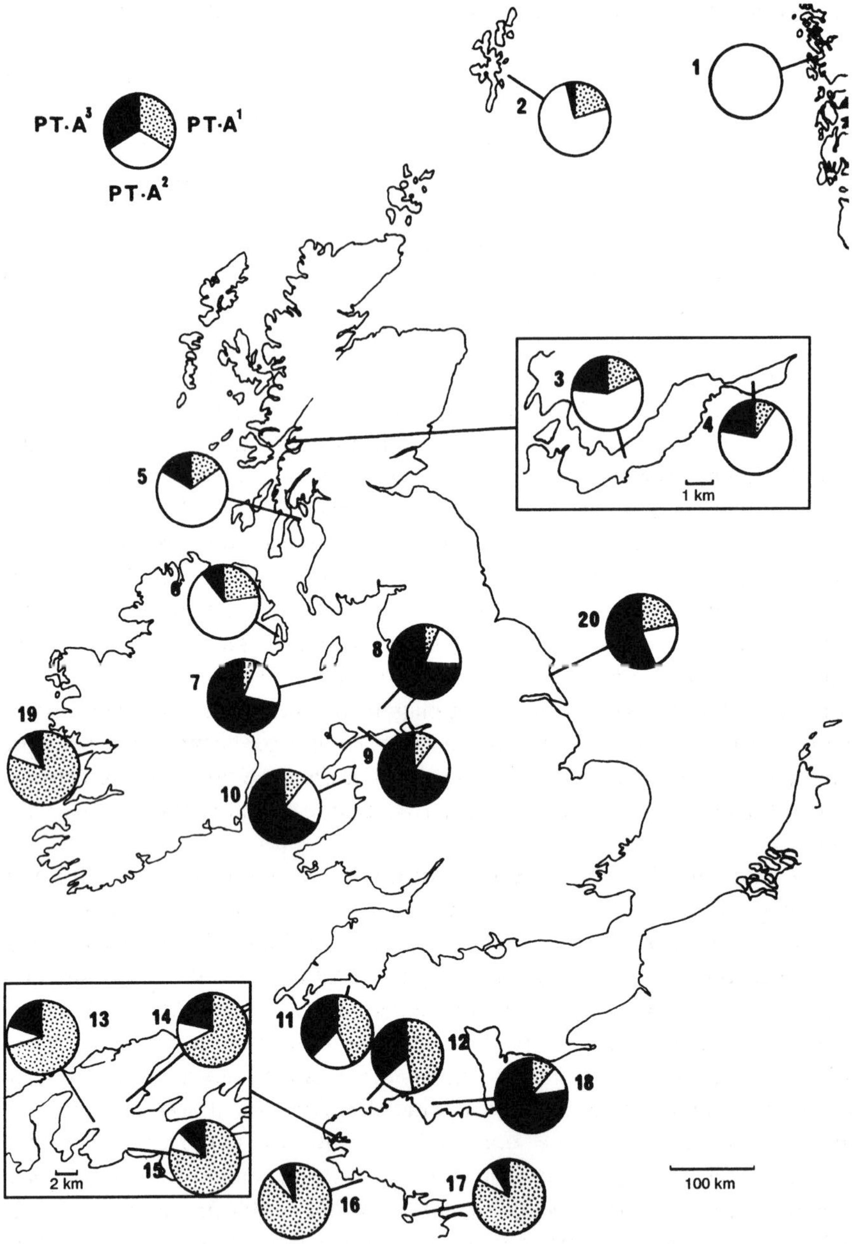

Fig. 3.2 Variation at a protein locus (*PT-A*) in the queen scallop (*Aequipecten (Chlamys) opercularis*) in northern European waters. Frequencies of the three alleles $PT\text{-}A^1$, $PT\text{-}A^2$ and $PT\text{-}A^3$ are given. Sample sites: 1 = Bergen, 2 = Shetland, 3 & 4 = Loch Creran, 5 = Firth of Clyde, 6 = Strangford Lough, 7, 8 & 9 = Irish Sea, 10 = Cardigan Bay, 11 = Plymouth, 12 = Roscoff, 13, 14 & 15 = Rade de Brest, 16 = Concarneau 17 = Belle Ile 18 = St Brieuc, 19 = Galway, 20 = Bridlington. (From Beaumont, A.R. (1982) Geographic variation in allele frequencies at three loci in *Chlamys opercularis* from Norway to the Brittany coast. *Journal of the Marine Biological Association, UK,* **62**, 243–261. Cambridge University Press.)

of the year and which could prevent the passage of larvae out of, or into, this body of water.

Accepting the possibility of restricted larval exchange due to hydrographic factors, how long could these populations have been isolated and is it long enough for random genetic drift to have produced such significant differences in allele frequencies? We do know that most of the Irish Sea was covered with ice during the last glaciation which ended 10 000 to 13 000 years ago. Therefore, this is the maximum time in which genetic drift could have created these differences. Current wisdom, based on Nei's 'molecular clock' (Box 3.3), is that this is not long enough.

How then do we explain this apparent genetic differentiation? Could there be restricted larval exchange together with selection? This is not an easy question to answer. There is no evidence for significant deviations from the Hardy–Weinberg model at this locus in any of the populations studied, so clearly any selection operating is not particularly strong. In order to demonstrate selection, weak or otherwise, we would need to know the function of the protein coded for at the *PT-A* locus and link this in some clear way through its biochemistry and physiology to some environmental factor which was different in the different populations. In this case, information on the protein function or any environmental link is not known. Another possibility to consider is that the *PT-A* locus might be located very close on the chromosome to another locus which is under strong selective pressure. In this case the *PT-A* locus might only be a marker for the selected locus.

From the point of view of fisheries management, the safest conclusion that can be drawn from the *PT-A* story is that there are, indeed, significant differences between certain populations of the queen scallop and that they are probably the result of limited larval exchange combined with weak selection for one or other allele in different regions.

Now we can consider a clear example of how allozyme data can be shown to send a misleading message. The American oyster, *Crassostrea virginica*, occurs in estuarine habitats throughout the east coast of the USA, from Maine in the north to the Gulf of Mexico in the south. It has a larval phase lasting several weeks and might be expected to be relatively genetically homogeneous as a result of this. Allozyme studies detected no significant genetic differentiation into local populations anywhere in the species' range. However, mtDNA markers and some neutral, non-coding nuclear DNA markers demonstrated a very distinct division into a southern group and a northern group of populations separated at a zone of demarcation near to Cape Canaveral on the Florida Atlantic coast (Fig. 3.3). This evidence makes it clear that there is practically no oyster gene flow between the southern and northern regions, probably owing to the characteristics of a major oceanic current – the Gulf Stream – which emerges from the Gulf of Mexico around the tip of Florida. On the basis that evolution is faster in the mtDNA molecule than at protein coding loci, it could be suggested that, although differences in the mtDNA molecule have evolved, insufficient time has elapsed since the divergence of the two groups of oysters for allozymes to reflect the division. However, the non-allozymic single copy nuclear DNA markers also show this divergence and the most convincing conclusion is that selection at the allozyme loci is maintaining the same alleles at similar frequencies on both sides of

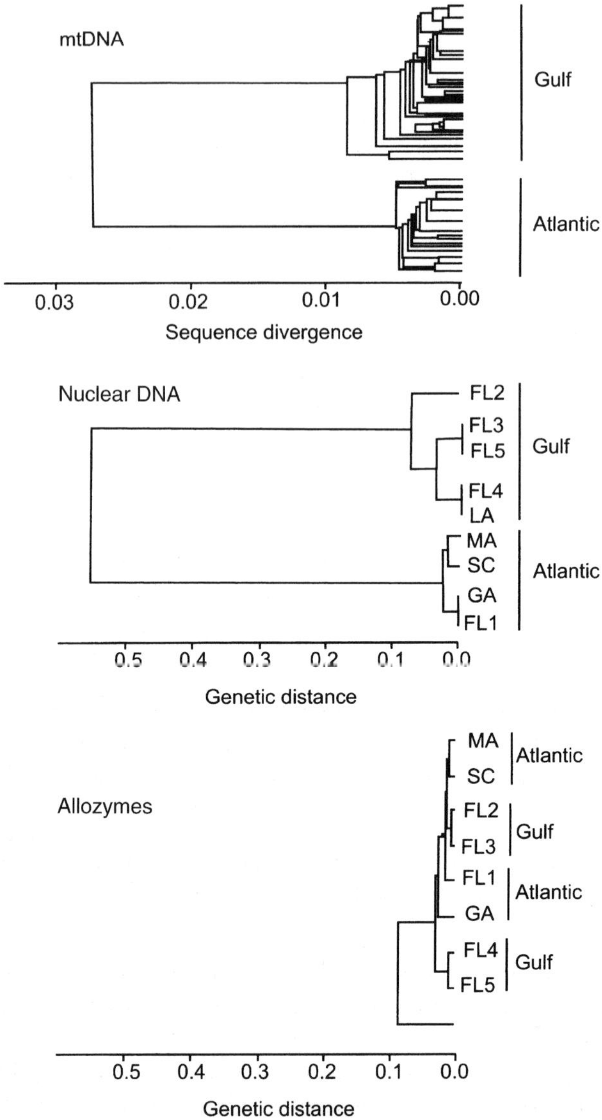

Fig. 3.3 Dendrograms illustrating estimated genetic relatedness between populations of the eastern oyster (*Crassostrea virginica*) along the east coast of the USA, based on genetic distances calculated from allozymes, nuclear DNA markers and sequence divergence in mitochondrial (mt) DNA. (Reprinted with permission from Karl, S.A. & Avise, J.C. (1992) Balancing selection at allozyme loci in oysters: implications from nuclear RFLPs. *Science*, **256**, 100–102. Copyright 1992. American Association for the Advancement of Science.)

the break point. It is notable that several other species such as the horseshoe crab (*Limulus polyphenus*), the black sea bass (*Centropristis striata*) and the seaside sparrow (*Ammodramus maritimus*) also exhibit a fundamental mtDNA divergence in the same region, and this is congruent with the biogeography of the region.

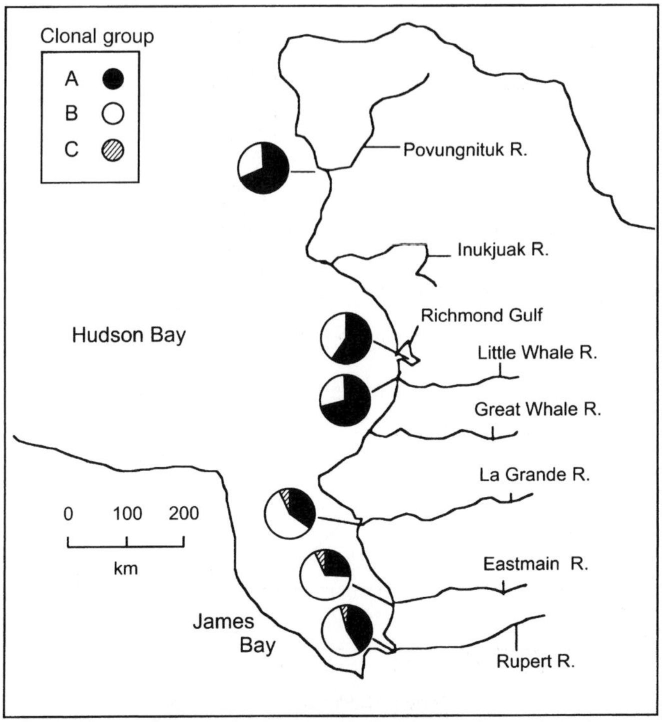

Fig. 3.4 Mitochondrial DNA markers used to identify genetic differentiation in cisco (*Coregonus artedii*) populations from rivers in the James and Hudson Bays in Canada. The clonal groups A, B and C have been identified following cluster analysis of cisco haplotypes. (Reprinted with permission from Bernatchez, L. & Dodson, J.J. (1990) Mitochondrial variation among anadromous populations of cisco (*Coregonus artedii*) as revealed by restriction analysis. *Canadian Journal of Fisheries and Aquatic Science*, **47**, 533–543.)

What then are we saying here? Is it really the case that allozyme data should not be trusted at all? This is a vexed question, but the answer is that it is probably safe to accept allozyme data when a large number of loci have been included. Fortunately, more rapidly evolving, selectively neutral markers such as mitochondrial DNA and microsatellites are now the tools of choice for studies of population differentiation.

An example of the use of mitochondrial DNA variation to detect population structure is given in Figure 3.4. The anadromous fish *Coregonus artedii*, the cisco, is important in the local fisheries of the river systems of the James and Hudson Bays in Canada. Allozymes were just not variable enough to show any differences between local river populations, therefore a mitochondrial DNA approach was used. Figure 3.4 shows the frequencies of three mtDNA clonal groups (A, B and C) of cisco from rivers feeding the Hudson and James Bays. The clonal groups are not individual mtDNA haplotypes, but come from three clear clusters on a dendrogram of the 19 haplotypes detected from the 141 fish sampled in the study. Different mtDNA haplotypes can be compared (based on presence or absence of cut sites for different restriction enzymes, see Box 3.4) to estimate their relatedness to one another and this

Box 3.4 Genetic distance measures based on DNA restriction fragments or DNA sequences

Restriction fragment length polymorphism (RFLP) data are derived from comparisons of DNA fragment sizes following cutting with restriction enzymes. The comparison is based on the proportion of shared fragments in the restriction digest products of two taxa and can be extrapolated to estimate the proportion of base substitutions, thus providing a value for sequence divergence.

The proportion of shared fragments (F) is given by:

$$F = 2N_{xy} / (N_x + N_y)$$

where N_x, N_y and N_{xy} are the number of restriction fragments observed in sequences X and Y and shared by X and Y respectively. The number of base substitutions per nucleotide (p) is then given by:

$$p = 1 - [0.5\,(-F + (F^2 + 8F)^{0.5})]^{\,1/r}$$

where r is the number of base pairs in the recognition site of the enzyme. Comparisons usually involve several restriction enzymes and a weighted mean of p is produced after first calculating p for each of the values of r (four-, five- or six-base cutters).

The sequence divergence can also be estimated by using shared 'sites' rather than shared fragments. Let N_x, N_y and N_{xy} now be the number of sites observed in sequences X and Y and shared by X and Y respectively. The proportion of shared sites S will then be:

$$S = 2N_{xy} / (N_x + N_y)$$

and the number of base substitutions is estimated by the simpler formula:

$$p = -\ln S/r$$

where r = the number of base pairs in the recognition site of the enzyme. A weighted mean of p is again produced by first calculating p for each value of r.

Nucleotide sequence data are compared by aligning one against the other and checking for differences. Closely related taxa will have few differences and these differences are simple to evaluate, while for distantly related taxa larger differences will be evident which can even make alignment impossible. Where alignment is possible, the percentage sequence divergence (p) is simply:

$$p = z_d / z_t$$

where z_d is the number of nucleotides which differ between two sequences and z_t is the total number of nucleotides in the sequence. For more distantly related taxa adjustment is needed to take account of nucleotides that may have undergone more than one mutation since the taxa diverged (multiple hits) and to accommodate the fact that transitions (purine–purine or pyrimidine–pyrimidine mutations) may be more frequent than transversions (purine–pyrimidine mutations or vice-versa).

Note that RFLP or sequence data from mitochondrial DNA effectively relate to haploid clones and can be used, like allele frequency based indices (Box 3.3), as measures of differences between individuals or taxa. The same may not be true of RFLP data or sequences from the nuclear genome where the differences observed may be between two alleles at a locus.

is what provides clusters of more closely-related haplotypes. The genetic pattern revealed by a detailed mtDNA analysis of the different haplotypes within the clonal groups suggests that the cisco of this region, which was covered by ice in the last glaciation, were actually derived from two quite separate glacial refugia (clonal groups A and B) and have recolonised the area by different routes. Since recolonisation, there has been some merging between the two types but this has not obscured the historical geographic pattern. The source of clonal group C which is strongly diverged from both the A and B clonal groups is not clear – it might even represent introgressed mtDNA from a closely-related hybridising species. In this example, we see how mtDNA analysis can not only tell us about the current population structure of a species, but also can point to historical processes which have influenced the pattern observed in the present day. In human genetics, this valuable characteristic of mtDNA has been used to provide evidence for the 'out of Africa' theory, since it appears that we are all descended from a 'mitochondrial Eve' who lived in Africa some 140 000 to 290 000 years ago.

The rate of evolution of microsatellites is much faster than for allozymes or mtDNA and therefore microsatellite variation is a more sensitive measure of weak genetic differences between populations. Figure 3.5 illustrates how much genetic differentiation can be resolved in salmon (*Salmo salar*) populations by microsatellites compared with allozymes. At the top of the figure, the dendrogram (Box 3.3) is based on an allele frequency analysis of 38 allozyme loci in populations from 31 rivers that drain into three main regions – the west Atlantic, the east Atlantic and the Baltic Sea. In the lower dendrogram equivalent results based on 4 microsatellite loci are displayed. You can see that the microsatellite genetic distances between river populations in the west Atlantic (D ranges from 0.06 to 0.11) are actually greater than the allozyme genetic distances when comparing salmon populations from both sides of the Atlantic Ocean ($D = 0.04$). So the power of microsatellites to identify population differentiation is very high. However, there are two caveats that must be added here. First, the resolving power of loci with many alleles, such as microsatellites, is highly dependent on sample size (Box 3.5). A small sample of, say, 10 individuals scored at a locus with 25 alleles will inevitably only be sampling a few of those alleles. Comparing this with another small sample from a different population that also has the same alleles at similar frequencies could show quite large genetic differences based on chance alone. Some alleles sampled from one population will be absent in the sample from the other population, leading to the mistaken conclusion that the two populations are very different. Secondly, Nei's genetic distance measure (D) was devised on the basis of estimated point mutation rates at allozyme loci (around 10^{-7} mutations per gene per

Box 3.5 Statistical problems associated with population genetic analyses

Sophisticated statistical analyses are beyond the scope of this book but there are a number of statistical problems or methods encountered in the analysis of genetic data of which the reader should be aware.

Degrees of freedom in goodness-of-fit tests of the Hardy–Weinberg model

Where goodness-of-fit tests are used to compare genotype frequencies at a locus against the Hardy–Weinberg model, the number of degrees of freedom in the test are less than might be expected. This is because observed genotype frequencies are first used to calculate allele frequencies, from which the expected genotype frequencies are then predicted. The formula for the number of degrees of freedom (d.f.) is:

$$\text{d.f.} = 0.5\,(n^2 - n)$$

where n is the number of different alleles at the locus.

Problems with sample size

When measuring population differentiation, how many individuals should be sampled from each potential population? This actually depends on the particular markers being used because the more variable the marker is, the larger must be the sample size. Imagine dipping into two bags of coloured sweets each with the same 12 colours and taking just five sweets from each bag. You would be likely to get a different set of colours from each bag and this might lead you to believe that the colours were different in the two bags.

Allozyme markers tend to be restricted to between 1 and 6 alleles per locus. In this case a sample of between 30 and 50 individuals is likely to pick up all the alleles and the full variation at the loci will be sampled. On the other hand, microsatellite loci often exhibit from 10 to 50 alleles. At such loci a sample of 30 individuals is unlikely to pick up all the variation. In addition, problems could arise in analysis because the allele frequencies in the two samples might be different on the basis of chance alone. The alleles and their frequencies could be identical in both populations, but limited sampling might suggest that they were different.

Problems with Type I statistical error: Bonferroni correction

In statistics we can sometimes erroneously conclude that something is significantly different from something else when actually it is not. This is called a

Type I error. If, for example you toss a coin 10 times, you would not expect to get nine heads and one tail. If you did, you might conclude that the coin was biased. However, if you carried out this trial 100 times, then you might expect to get this result maybe once or twice. In this instance, you would not suspect the coin to be biased. Why? Because we intuitively understand the concept of probability – if something is unlikely, it has a low probability and it will happen rarely. In the coin example, the hypothesis that is being tested is that the coin is unbiased – it is as likely to fall heads as tails. If we test that hypothesis hundreds of times, the improbable outcomes (such as 10 heads, or 10 tails) will eventually occur. But the coin is never biased and the hypothesis remains true. Now extend this idea to tests of agreement to the Hardy–Weinberg model at say 10 allozyme loci in perhaps 12 different populations – the type of situations often encountered in population genetic surveys. Here we have 120 tests of the same hypothesis (that genotype frequencies will not differ from the Hardy–Weinberg model) and perhaps we find that 10 of the tests show significant differences from the model. Are we to conclude that the hypothesis is untrue in these 10 instances? Or could this just be the result of chance? Fortunately there are simple methods of adjustment of significance levels called Bonferroni corrections which allow for these instances of chance so that we can be more sure which of the 10 test results really are genuinely significant.

Are genetic indices designed for gene coding loci suitable for microsatellite loci?

Many genetic indices, such as Wright's F-statistics and Nei's D, were developed based on the assumption, among others, that variation is due to point mutations in the DNA sequence and that there is a relatively constant rate of point mutations. In addition, there is an assumption that all new alleles arise *de novo* (one allele does not change directly by one point mutation back to an existing allele) and that there are an infinite number of potential alleles at a locus. This is the infinite allele model (IAM). These assumptions are reasonable for allozyme loci, but have been criticised as assumptions for microsatellite loci. The reason for this is that microsatellite alleles differ from one another by the number of repeated elements they contain. Mutations giving rise to new microsatellite alleles are not point mutations, but are mutations due to slippage during replication. There are not an infinite number of new alleles (the length of microsatellites sequences does seem to be controlled in some way) and the expectation is that one allele gives rise to another by the addition or loss of a repeated element. This is called the stepwise mutation model and a number of geneticists have devised alternatives to F_{ST} (e.g. the statistic R_{ST}) and other indices to allow for this difference in the evolution of alleles at microsatellite loci. Comparisons between the suitability of F_{ST} and R_{ST} for use with microsatellite loci have been made, but this issue has yet to be resolved.

individual). As mutation in microsatellites is due to slippage during replication (around 10^{-3} mutations per gene per individual) rather than point mutation, it would not be safe to use Nei's formula (time of divergence = $5 \times 10^6 D$; Box 3.3) to extrapolate any times of divergence between taxa based on microsatellite data (Box 3.5).

Mixed stock analysis (MSA)

When an oceanic salmonid catch is brought ashore it probably consists of fish from a mixture of distinct populations from different river sources. We have already seen how salmonids can be genetically very divergent between rivers and these differences are likely to be reflected in factors such as growth rates, age at maturity or other life history parameters. If the fishery is managed as a single unit there is the potential for the overexploitation of fish from the less abundant populations, or the suboptimal harvesting of fish from those that are more abundant. In order to enable proper management and obtain the maximum sustainable yield from such fisheries, we need to estimate the proportions of the various populations in the fished resource. If the contribution of the various populations varies geographically or over time, then fishing effort can be directed to exploit the strongest populations and pressure on weaker populations can be reduced by targeting fishing to particular regions or particular times.

Early attempts to tag or brand fish as they departed from specific rivers were relatively ineffective because of the high initial labour costs and the loss of markers in a proportion of the fish. Far more effective is the use of genetic markers. Allozyme markers have been widely used in the management of Pacific salmon based on comparisons between allele frequency data for the individual river populations and allele frequencies of the total catch. Of course, to make it work, the differences between river populations must be sufficiently great that allele frequencies at, minimally, a few of the loci scored will enable identification of that population in the total catch. Fortunately, there are statistical methods available which make these comparisons possible even when allozyme–allele frequency differences are not very great. The potential resolving power of microsatellite markers is much greater than allozymes (see Fig. 3.5) and it is expected that they will be used increasingly in MSA.

MSA is also relevant to some sport fisheries that consist of a mixture of natural and introduced stock. The introduced element is either transplanted from another area or is derived from hatchery production. A good example is the trout, *Salmo trutta*, which exists in Europe in two different forms. One, the resident or brown trout, inhabits rivers and lakes and never goes to sea, while the other, the sea trout, is anadromous and, as its name suggests, migrates into the sea and spends some time there before returning to its natal river to spawn. Where waterfalls that are impassable to trout are present in rivers, the populations above the falls will be resident trout and those below will be sea trout. Because sea trout grow bigger than resident trout they make a better fighting fish for rod and line and, mainly for this reason, restocking of river populations above impassable falls is often carried out with hatchery-reared sea trout. The expectation was that there would not be a problem with the

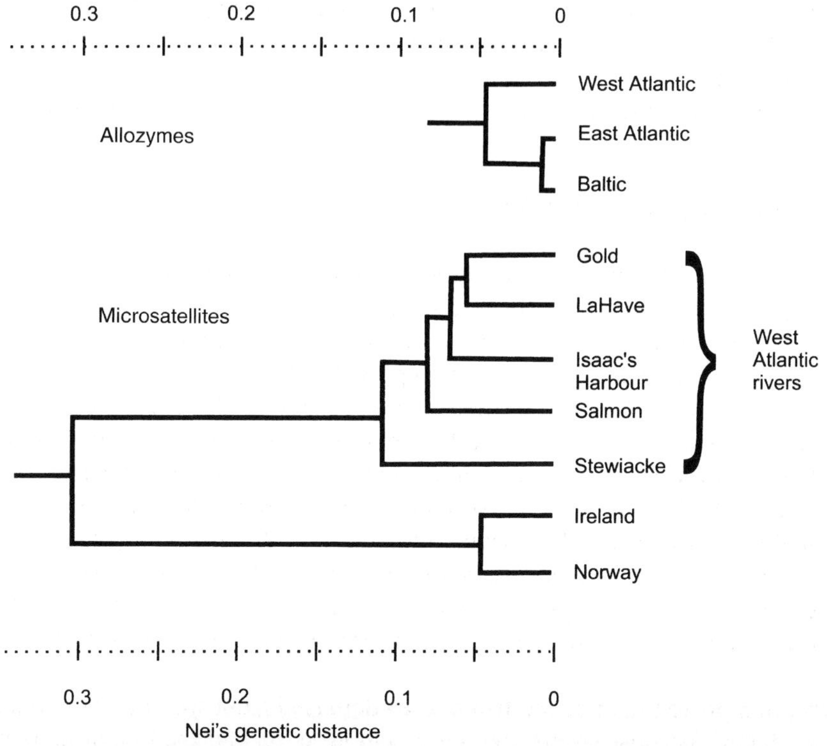

Fig. 3.5 A comparison of genetic variation in Atlantic salmon (*Salmo salar*) populations in localised and broad geographic regions using allozymes and microsatellites. (Modified from McConnell, S., Hamilton, L., Morris, D., Cook, D., Paquet, D., Bentzen, P. & Wright, J. (1995) Isolation of salmonid microsatellite loci and their application to the population genetics of Canadian east coast stocks of Atlantic salmon. *Aquaculture*, **137**, 19–30. Copyright 1995, with permission from Elsevier Science.)

survivors of these introduced sea trout breeding with the natural population of resident trout because, once they had migrated to the sea, they would not be able to return to their original position (above impassable falls) in the river to breed. However, allozyme-based studies in various parts of Europe have shown that this does not always happen. It is clear that some stocked sea trout do not migrate to sea but remain in the populations of resident trout with which they will then breed. This leads to a loss of the unique genetic identity of the resident stocks and is an important conservation issue.

In North America the cutthroat trout exists as a complex of up to 15 different recognised subspecies and there has been extensive introduction of the different subspecies into non-native habitats. Allozyme analysis has demonstrated that this has resulted in an extremely complex pattern of hybridisation and introgression. This is bad news because the hybrids often suffer developmental abnormalities that make them less fit than the pure-bred subspecies. Furthermore, as in the case of brown trout in Europe, there is evidence of genetic swamping which threatens the uniqueness of locally-adapted populations.

Conservation genetics

Most educated people are now aware of the fragile state of the planet and the increasing pressures from human activities on the animals and plants with which we share the biosphere. Species are becoming extinct at a rate comparable to the mass extinctions of geological time and, in addition to loss of species, there is a loss of biodiversity within remaining taxa. In this case, for 'conservation of biodiversity' read 'conservation of genetic diversity'. How then do we measure genetic diversity in a species and, having established that genetic diversity is reduced, what can we do about it? The commonest measure of genetic diversity is heterozygosity: at a single locus this is the proportion of individuals that are heterozygous and this value can be averaged across as many loci as there are data for. Another measure of diversity is the average number of alleles present across all loci. These measures of diversity are discussed in more detail in the next chapter, where we deal with genetic problems in the hatchery.

One of the difficulties in assessing genetic diversity is that many marine species exhibit the phenomenon of chaotic patchiness. This is the situation where there is extensive microspatial variation in allele frequencies detected at any one sampling time, but allele frequency changes occur over time such that the pattern observed might be very different if sampled at another time. Reproductive strategies in the marine environment are likely to create chaotic patchiness because of the explosive reproductive capacity of just a few individuals. Therefore, the lucky survivors of spawning and larval development are not always the average genetic representatives of the parent population. Fortunately, in the case of commercial marine species – probably because they normally have very large population sizes – studies have tended to show little significant temporal variation in allele frequencies. The problem really arises and becomes acute where population sizes have for some reason (e.g. fishing, habitat loss, disease) become much smaller. When population numbers dwindle to worrying scarcity and the species could become extinct, we have to consider how to manage the conservation of the species actively. Many of our fish species have experienced dramatic declines (cod and herring are good examples), but at what point are conservation measures required to prevent extinction?

Allele numbers and heterozygosity can be assessed using the techniques described in Chapter 2 and this would be the first step in any programme to assess the extent of genetic diversity in a threatened species. Beyond this, there is little that geneticists can do apart from lobbying the appropriate authorities in an attempt to alleviate the pressure on the species. If the situation is so dire that the numbers of individuals in a species fall to very low levels then the problem becomes graver because of the effect of inbreeding depression (Chapter 4, Box 4.1). Artificial breeding programmes are then required and in Chapter 5 we show how breeding plans can be developed to maintain the greatest genetic diversity in spite of small population size.

Considering marine fish and shellfish species, complete extinction from overfishing is an unlikely outcome because fishing pressure is usually reduced when it becomes uneconomic to target a particular species. There could still be sufficient numbers of individuals of that species out there to maintain a small but viable population. The type of fish species which are more vulnerable to extinction from overfishing or habitat loss are those which have colonised lakes following the retreat of the ice sheets of the last glaciation. Such species are often strongly substructured, with particular subspecies or variants restricted to just one or a few lakes. Finally, we must mention marine mammals, which have been probably the highest profile group of threatened species in the marine environment. It is difficult to secure samples for genetic study from whales. Indeed, until the development of DNA techniques which required only minute quantities of tissue, such as can be obtained from the sloughed-off skin, little genetic information was known about cetaceans. More recently, using mtDNA, microsatellites and DNA fingerprinting, together with observations of behaviour in the wild, much has been learned about the family relationships between individuals within pods. Such information has assisted the conservation effort, not least by revealing and publicising the complicated and subtle nature of the population substructure of these species.

In addition to their value in addressing the wider questions of population structure, molecular genetic markers can be of great forensic importance. Naturally, legislation to protect threatened species or to manage fisheries relies on correct identification of an organism or parts of it. In the case of many fish and shellfish it may often only be a piece of muscle tissue from which a prosecution must be mounted. There are many examples of the use of allozyme markers to identify fish species in order to prove cases of illegal fishing, but here we will consider the problem of identification of two commercially important species of scallop in the North Atlantic. The large sea scallop (*Placopecten magellanicus*) fishery is of major importance on the Atlantic coast of Canada and the USA but, in the same region and extending across the northern Atlantic, there is a fishery for the smaller Iceland scallop (*Chlamys islandica*). Scallops for the American market are shucked (removed from the shell) at sea with only the adductor muscle (meat) being landed and a critical management tool for regulation of these fisheries is to use meat counts. Given that the meats look the same, it would be easy for fishers to pass off small, undersized sea scallop meats as Iceland scallop meats. Although meats are often up to 12 days old by the time they are landed, identification of adductor muscle tissue of the two species has been achieved using two allozyme loci that are diagnostic. In addition, diagnostic markers have been established by RFLP analysis and by direct sequencing of a portion of the 18S rRNA gene. Thus, molecular biology has provided fisheries managers with simple, relatively cheap and effective tools to assist with the enforcement of legislation.

Further reading

Avise, J.C. (1994) *Molecular Markers, Natural History and Evolution*. Chapman & Hall, London.

Beaumont, A.R. (ed.) (1994) *Genetics and Evolution of Aquatic Organisms*. Chapman & Hall, London.

Carvalho, G.R. (ed.) (1998) *Advances in Molecular Ecology*. IOS Press, Amsterdam.

Carvalho, G.R. & Pitcher, T.J. (eds) (1995) *Molecular Genetics in Fisheries*. Chapman & Hall, London.

Hillis, D.M. & Moritz, C. (eds) (1996) *Molecular systematics*, 2nd edn. Sinauer Associates Inc., Sunderland

Kenchington, E., Naidu, K.S., Roddick, D.L. Cook, D.I. & Zouros, E. (1993) Use of biochemical genetic markers to discriminate between adductor muscles of the sea scallop (*Placopecten magellanicus*) and the Iceland scallop (*Chlamys islandicus*). *Canadian Journal of Fisheries and Aquatic Science*, **50**, 1222–1228.

Ryman, N. & Utter, F. (eds) (1987) *Population Genetics and Fishery Management.* University of Washington Press, Washington.

Ward, R.D., Skibinski, D.O.F. & Woodwark, M. (1992) Protein heterozygosity, protein structure and taxonomic differentiation. *Evolutionary Biology*, **26**, 73–159.

Chapter 4
Genetic Considerations in the Hatchery

In the previous chapter we covered genetic aspects of wild populations of fish and shellfish. Now we need to consider how those genetic principles might be important in the hatchery environment. The principal function of a fish or shellfish hatchery is to produce large numbers of juvenile or market-sized individuals economically. For this purpose a broodstock of adults is initially collected from the wild. This broodstock may be renewed from the wild each season, or it may be retained and added to, over the years, by selected hatchery-reared individuals. The importance of artificial selection of fish and shellfish and the methods employed are dealt with in Chapter 5, but the uncontrolled use of closely related individuals such as fathers, daughters, mothers and sons in a broodstock will lead to inbreeding. From the genetic perspective, inbreeding increases homozygosity (Box 4.1) and almost always has deleterious phenotypic effects such as reduced larval and juvenile performance or viability. These phenotypic effects of inbreeding are called inbreeding depression. In Chapter 5 we will discuss how breeding plans need to take the dangers of inbreeding into account.

Although inbreeding itself will generally reduce the performance of offspring, the process of inbreeding does have a potential value. It can be used to develop highly homozygous 'lines' of aquacultural animals. Such lines will be homozygous at most of their gene loci, but different inbred lines will be homozygous for different alleles at many of the loci. When two separate inbred lines have been produced, these can be crossed with one another to produce 'F1 hybrid' offspring which will then be heterozygous at these loci. F1 hybrids are expected to demonstrate above average performance (hybrid vigour, discussed below) and in agricultural animals and crops this approach has proven to be extremely effective. However, apart from the specific value of F1 hybrids, the first important genetic consideration for the broodstock is to avoid or reduce inbreeding.

The second, but equally important, consideration is that broodstocks consist of relatively few individuals compared with the size of the population in the wild. In earlier chapters we have emphasised the boundless quantities of genetic variation present in natural populations. However, if you take just a few (say, fewer than one hundred) individuals for your broodstock then much of the natural genetic variation will be missing. Does this matter and, if so, how much?

Each variant allele at each coding locus in a population can be regarded as part of the 'genetic resource' of that population. An allele alone, or in combination with other alleles or loci, could be responsible for conferring on its carrier a valuable trait such as increased resistance to a particular disease, better cold tolerance or faster growth. Therefore, the loss of any allelic variants is a potential loss of valuable genetic resource. Of course, if most allelic variation at coding loci is neutral, then this is less important, but we would be unwise to ignore the certainty that at least *some* variants at coding loci will be advantageous. If not now, then most likely in the near- to

Box 4.1 Inbreeding

Inbreeding is mating between related individuals. The more closely related two individuals are, the greater the level of inbreeding. Brother and sister matings are therefore more strongly inbred than, for example, matings between cousins. The most extreme form of inbreeding is self-fertilisation.

In a homozygote, the two alleles present could have arisen by replication of a single allele from a direct ancestor – for instance, an allele could be passed from an individual to two offspring that then breed together, so that some of their offspring inherit two copies of the allele. Such alleles are described as identical by descent, or IBD. Alleles that are alike but which are not recent copies of ancestral alleles are said to be alike in state, or AIS. We assume a starting population in which all homozygotes are between AIS alleles – this population is non-inbred. An inbred population, however, will have homozygotes with AIS alleles and also a certain proportion of homozygotes with IBD alleles. The inbreeding coefficient (F) is the probability that the two alleles at a locus are IBD and is a measure of increased homozygosity against the Hardy–Weinberg model (Chapter 3, Box 3.1). $F = 0$ in the starting or reference population where there are no IBD homozygotes and $F = 1$ where there is complete IBD homozygosity within the population. The degree of inbreeding experienced by particular individuals can be estimated from an analysis of their pedigree; for example, the probability of IBD alleles in the offspring of siblings is 1:4 ($F = 0.25$) and in the offspring of half-siblings is 1:8 ($F = 0.125$).

To illustrate how inbreeding causes an increase in homozygosity (or equivalently, a reduction in heterozygosity) we can take the example of self-fertilisation in hermaphrodites which is the extreme form of inbreeding (Fig. B4.1). After

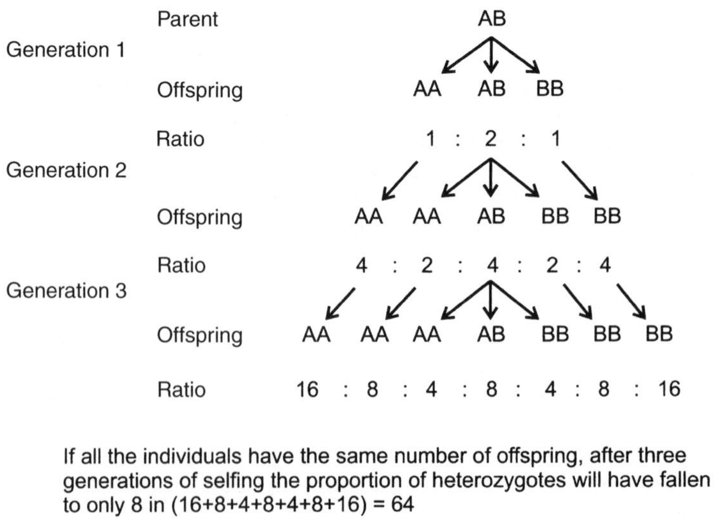

Fig. B4.1 Illustration of the loss of heterozygosity following several generations of self-fertilisation. A and B are alleles at a locus.

one generation of self-fertilisation, half of the resulting offspring will be IBD homozygotes. The proportion of heterozygotes halves with each generation of selfing, so, if all individuals have the same number of offspring, after three generations of self-fertilisation the proportion of heterozygotes will have fallen to only one in eight, that is, IBD will have risen to 87.5%.

medium-term future, global warming will bring about the increasing importance of high temperature resistant allelic variants at biochemically important loci in temperate aquaculture species. Such alleles may be effectively neutral until extreme summer temperatures reveal their value.

Is there evidence of loss of genetic variation in the hatchery?

Although the example used in this chapter concerns oysters, the principles are the same for all aquaculture species. In the early and mid-1900s, natural fisheries for the European flat oyster (*Ostrea edulis*) and the American oyster (*Crassostrea virginica*) were in serious decline. In order to meet the demand for oysters, the Pacific or Japanese oyster (*Crassostrea gigas*) was imported from Japan to several areas of the world and this species soon became the most important cultured oyster species worldwide. Several early importations of Pacific oysters were made to British Columbia on the Canadian west coast and the Pacific oyster became naturalised in that region. This naturalised population was a major source of broodstock for hatcheries along the west coast of the USA. In spite of the reduction in abundance of flat oyster (*O. edulis*) in Europe, there remain some natural populations and small-scale hatchery production of this species continues. Table 4.1 illustrates how the mean number

Table 4.1 Loss of genetic variation in hatchery populations. Pacific oyster (*Crassostrea gigas*) data based on 14 allozyme loci after three generations of separate hatchery breeding involving hundreds of broodstock oysters. Flat oyster (*Ostrea edulis*) data from four microsatellite loci on 10 populations of wild oysters and two populations derived from a hatchery

	Mean number of alleles per locus	Observed heterozygosity	Estimate of N_e
Pacific oyster – allozymes			
Wild	3.50	0.262	
Hatchery 1	3.07	0.268	40.6
Hatchery 2	2.43	0.256	8.9
European flat oyster – microsatellites			
Wild population mean	17.49	0.843	
Hatchery population mean	9.15	0.782	

(Data adapted from Hedgecock & Sly,1990 and Beaumont & Sobolewska, unpublished.)

of alleles per locus has been reduced as a result of hatchery activities in both species. The average number of alleles at the microsatellite loci is much higher than for the allozyme loci and therefore the reduction in the number of alleles due to hatchery activity is much more dramatic.

Assuming random genetic drift to be the main cause of genetic divergence, the effective population size of the hatchery-produced Pacific oysters could be calculated. Even though hundreds of broodstock were used, it can be seen from Table 4.1 that effective population sizes are estimated at around 40 individuals for one hatchery and fewer than 10 for the other. These numbers are much smaller than would be expected and imply that even though many oysters might have been used in the spawnings, only a proportion of them make any contribution towards the final cohort of offspring. Why might this be?

The Pacific oyster is amenable to strip-spawning and this method of cutting the oyster open and teasing out the gametes is used in almost all commercial hatcheries. Therefore, there is no uncertainty about the numbers of individuals providing gametes for a mass spawning. On the other hand, there is no guarantee that stripped gametes from all individuals are fully ripe and able to fertilise. This could be one source of the discrepancy between the estimated and the actual number of spawners. Another source of this discrepancy is applicable to other oysters and most other bivalves that are not amenable to strip spawning and thus have to be induced to release their gametes by some chemical or physical stimulus. For practical reasons, this induction is usually carried out with the broodstock held together in a large tank, and the water can become very milky with suspended eggs and sperm from just two or three individuals. All individuals will not necessarily respond to the stimulus and it is virtually impossible to tell whether there are more than just a few individuals spawning. Thus, this can reduce the effective numbers of broodstock. In the case of the flat oysters, which brood their eggs and larvae, control over spawning is even more difficult to achieve.

Once fertilisation has been achieved, it may be that some families perform far better than others. Slow growing families of oyster larvae will be lost from the system because of the size selection that takes place when sieve sizes are increased – only the larger larvae are retained. Alternatively, some families may just have a poor survival history and therefore will not be strongly represented among the hatchery-produced cohort. This is particularly relevant where, as in the example given (Pacific oysters, Table 4.1), each generation of broodstock is used to produce the next, as this is likely to result in inbreeding with concomitant slower growth and higher mortality.

In the case of many commercially important crustacean shellfish there are particular difficulties in the control of reproduction. Simple external fertilisation does not take place – females collect spermatophores from males and then at some later date use sperm from the spermatophores to fertilise the eggs, which they brood on specialised carrying legs. Putting equal numbers of males together in a tank with females is not always successful in ensuring good genetic mixing – in some species, such as *Macrobrachium rosenbergii*, the largest males may succeed in fertilising most of the females and prevent all other males from taking part.

This brings us to a separate difficulty: if there is an uneven contribution between males and females, how does this effect the situation? The formula to estimate effective population size in a spawning broodstock where different numbers of males and females are taking part is:

$$N_e = \frac{4\,N_f\,N_m}{N_f + N_m}$$

where N_e = effective population size, or effective number of spawners, N_f = number of females and N_m = number of males. So, if there are 20 individuals of which only three are male, then the effective number of spawners is only 204/20 = 10.2. Therefore, although 20 individuals are spawned, because of the imbalance between males and females, this is effectively equivalent to spawning only about five males and five females. Also, it must be remembered that it is the effective number of spawners that dictates how much of the genetic variation present in the original stock is transmitted into the offspring.

It can be seen that there are a number of explanations as to why, in spite of the best efforts of hatchery managers, the number of broodstock individuals which pass on their genetic variation to hatchery-produced offspring is likely to be fewer than predicted.

One practical solution to overcome some of these difficulties is to carry out several separate mini-spawnings between even numbers of males and females, and then combine the resulting embryos. This increases the certainty that the early larvae or fry are derived from as many parents as possible, but can do little to overcome the problems associated with differential growth or survival between families.

How do hatcheries affect heterozygosity?

So far we have considered how genetic variation measured by numbers of alleles can become depleted in hatcheries. Another measure of genetic variation – heterozygosity (the proportion of individuals that are heterozygous at a locus) – might also be expected to decrease. However, this does not appear to be the case for the oysters detailed in Table 4.1. Observed heterozygosity in the hatchery-derived oysters is not significantly lower than heterozygosity in the original populations of either species of oyster.

One explanation for the retention of heterozygosity is that it is not simply the number of alleles at a locus, but rather their relative frequency, that makes the strongest contribution to heterozygosity. Equal allele frequencies at a locus will give the highest heterozygosity and very unequal frequencies the lowest (Box 4.2). We can envisage that very unequal allele frequencies at a particular locus in a source population will tend to produce more equal allele frequencies in a hatchery population for the alleles which are, by chance, retained there. Thus, heterozygosity is not only retained, but possibly increased. However, in the case of loss of an allele at a bi-allelic locus there will be a dramatic loss of heterozygosity to zero. If many of the loci scored are

2 allelic variants @ a locus.

Box 4.2 The relationship between allele frequencies and heterozygosity

Let us consider two examples, using the formula for calculating expected heterozygosity at a locus:

$$\text{expected heterozygosity, He} = 1 - \Sigma x_i^2$$

where x_i is the frequency of the *i*th allele at the locus.

Example 1
A locus with four alleles, each present at an equal frequency of 0.25:

$$\text{heterozygosity} = 1 - (0.25^2 + 0.25^2 + 0.25^2 + 0.25^2) = 0.75$$

So, at this locus, 75% of the population should be heterozygotes.

Example 2
A locus with four alleles, one of which is common (frequency 0.85) and three of which are rare (each present at a frequency of 0.05):

$$\text{heterozygosity} = 1 - (0.85^2 + 0.05^2 + 0.05^2 + 0.05^2) = 0.27$$

In this case, although four alleles are present as in Example 1, the heterozygosity is much less because of the unequal allele frequencies.

bi-allelic, as is often the case for allozymes in fish and crustaceans, then reductions in numbers of alleles are more likely to be mirrored by reductions in heterozygosity. In fact, where similar studies have been carried out on fish and crustacean shellfish they have indeed generally shown reduced numbers of alleles and reduced heterozygosity. Studies on other bivalve shellfish have been similar to the oyster story (Table 4.1), with loss of alleles but little loss of heterozygosity.

Another factor which might account for some retention of heterozygosity in a bivalve hatchery is the common practice of artificial selection of larvae on the basis of size: small, slow-growing larvae are discarded during the process of sieving as pore sizes increase. How might this affect heterozygosity?

Geneticists are interested in heterozygosity in individual organisms, as well as in populations. We can ask the question: Is it advantageous for an individual to have two allelic variants expressed at a locus (heterozygosity) rather than having just one allele expressed (homozygosity)? From the point of view of the genome as a whole, the answer would seem to be clear. In Charles Darwin's famous book *On the Origin of Species by Means of Natural Selection* (1859), in which he lays out his theory of evolution, he states, '*I have collected so large a body of facts, and made so many experiments, showing, in accordance with the almost universal belief of breeders, that with animals and plants a cross between different varieties, or between individuals of the same variety but of another strain, gives vigour and fertility to the offspring, and, on the other hand, that close interbreeding diminishes vigour and fertility.*' Nothing discovered since

Box 4.3 The correlation between multiple-locus heterozygosity (MLH) and physiological parameters

Although there are many parameters that can be used to quantify aspects of fitness in an organism, one of the easiest to measure is size at age. This assumes that differences in growth rates of individuals of the same age, grown in the same environment, are a reflection of genetic differences between them. To test whether there is a relationship between size at age and genetic heterozygosity, individuals of the same age are weighed (or measured) and then scored at a number of genetic (usually allozyme) loci. Some will be heterozygous at, say, three of the loci (and homozygous at the rest), others heterozygous at four, five or six etc, loci. Each has a weight and a heterozygosity class to which it can be assigned and these can be graphed (Fig. B4.3) and tested to see if there is a significant relationship between them. Usually there is a weak positive correlation as illustrated.

One other feature which is commonly reported when tests of this type are carried out is the fact that variance around the mean weight is greatest in the low heterozygosity classes and decreases in high heterozygosity classes. So, although the average weight of highly homozygous oysters is low, there is a wide spread of weights among this group. In contrast, there is a narrow spread around the mean weight of the most highly heterozygous oysters. The suggestion is that the development and growth of highly heterozygous individuals is better determined and more tightly controlled than that of extensively homozygous individuals.

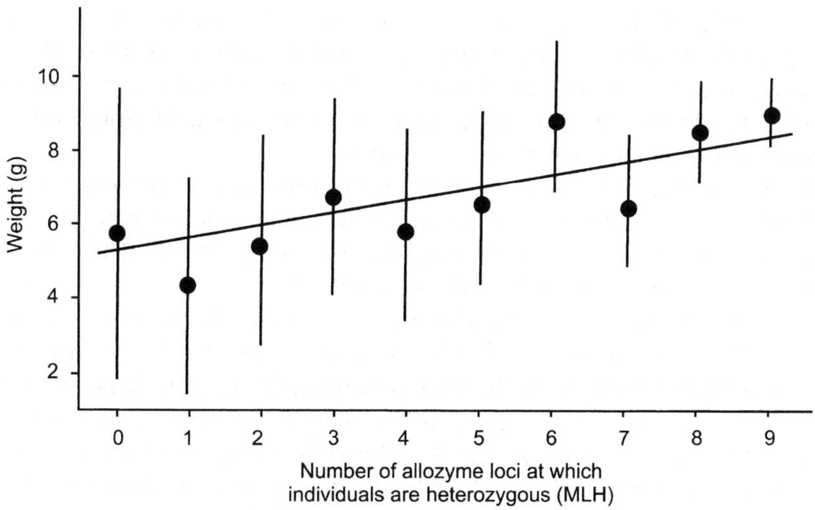

Fig. B4.3 Illustration of the principle of increased weight and reduced variance with increasing multi-locus allozyme heterozygosity (MLH). Vertical lines through points on the graph indicate the standard error of the mean. There is a weak (but significant) positive correlation between MLH and weight and the variance around the mean weights decreases with increased MLH.

This idea has been extended to test for control, during development, of bilaterally symmetrical features in fish such as fin ray numbers and fin dimensions. There is evidence (although weak and sometimes disputed) that highly homozygous fish (at a few allozyme loci) are more asymmetrical than those that are highly heterozygous. This phenomenon of fluctuating asymmetry has been looked at in humans too. According to some research, human facial symmetry is a very important fitness-related character that we use in selecting our partners. It is noticeable that catwalk models have highly symmetrical faces and bodies and, apparently, it is partly this symmetry that we use unconsciously to define attractiveness.

What are the mechanisms by which high heterozygosity could increase fitness and possibly reduce asymmetry? There have been two main theories. One theory is that the possession of two different alleles at many loci provides an individual with an advantage because biochemical differences between the products of the alleles enable them to deal with a wider biochemical environment both inside and outside the cell. The main alternative idea is that there is no particular advantage to being highly heterozygous but that there is a distinct disadvantage for an individual to be highly homozygous, because this exposes faulty alleles (deleterious recessive alleles) in the homozygous condition. It is beyond the scope of this volume to review the evidence one way or the other, but it is likely that both mechanisms operate across the genome as a whole.

Darwin's time has changed this view: inbreeding generally does reduce viability and performance, and hybrid vigour can often be consistently demonstrated. Knowledge of the genome enables us to associate the phenotypic characters of inbreeding depression with high homozygosity across the genome and the phenotypic characters of hybrid vigour with high heterozygosity across the genome. However, can this effect be detected in studies of just a few loci?

Perhaps surprisingly, there is now quite a lot of evidence to support the idea that multiple-locus heterozygosity (MLH) across just a few allozyme loci is positively (albeit weakly) correlated with size in juvenile bivalves (Box 4.3). Individuals that are heterozygous at many of the allozyme loci scored in genetic studies are, on average, larger than equivalent individuals of the same age that are homozygous at many of the scored loci. If this correlation is also present during the larval stage, then the loss of small larvae will increase the average heterozygosity of the remaining larval cohort.

So far in this chapter we have examined the genetic disadvantages of hatchery production, that is, the loss of genetic variation. However, for much of the rest of this volume we will look at how we can use molecular methods and other genetic techniques to advance aquaculture, beginning with the opportunities to genetically tag released organisms, to locate genes of interest and to tackle diseases.

How can we use genetic markers to identify hatchery-produced individuals?

There are a number of situations where being able to identify hatchery-produced individuals is of great importance. We have noted in previous comments about mixed stock analysis in salmonids that it is sometimes possible to use allozymes to identify certain natural stocks. There have been many cases of loss of hatchery-reared salmon into the wild, caused by breakage of nets in storms or due to boat damage. How can we identify these fish in order to find out what happens to them? Do they survive? Do they become part of the wild population? If they do, does this cause a genetic depauperisation of the wild stock?

Identification to family level

The use of a few highly variable microsatellite markers can make it possible to identify individuals to particular families. With enough variation at three or four loci, the multiple genotypes of individual broodstock parents will be unique to those parents and therefore to each hatchery-cross carried out.

Identification to population level

A good illustration of the use of an allozyme marker to detect hatchery-produced individuals among wild stock is provided by the work of Knut Jørstad and colleagues in the development of cod ranching in Norway. Cod is an extremely important commercial species in the North Atlantic and has suffered from overfishing to such an extent that the famous Grand Banks fishery grounds off Newfoundland and Labrador are now completely closed. Some level of artificial production of cod for enhancement of wild stocks was carried out in Norway over much of the twentieth century, but the evaluation of any release project is dependent on a reliable method for the identification of the released fish. Traditional identification methods involve mechanical or chemical tagging, but these can be time-consuming and expensive and they are not always readable in older fish.

Variation at the glucose phosphate isomerase (*GPI-1**) locus in wild populations of cod off Norway includes the rare *GPI-1* 30* allele. Using electrophoresis of tissue from fin-clips, individuals heterozygous for this rare allele were identified and spawned together. Yolksac larvae, 25% of which would be *GPI-1* 30/30* homozygotes, were reared in artificial seawater ponds and later, when mature, these *GPI-1* 30/30* homozygotes were identified from fin-clips and selected for spawning. Offspring of these matings, all *GPI-1* 30/30* homozygotes, were used in a release experiment in Masfjorden together with an equal number of a control group from matings between unselected cod. Juvenile fish from both groups were chemically tagged with oxytetracycline supplied in the diet so that they could be identified when

Table 4.2 *GPI-1** allele frequencies in populations of cod sampled from Masfjorden during an experiment to test the value of the rare *GPI-1* 30* allele as a marker for hatchery-produced fish

Fish sampled	*GPI-1* 30*	*GPI-1* 70*	*GPI-1* 100*	*GPI-1* 150*
4-year-old wild cod	0.03	0.00	0.71	0.26
3-year-old wild cod	0.03	0.00	0.71	0.26
2-year-old wild cod	0.03	0.03	0.66	0.28
1-year-old wild cod	0.05	0.01	0.75	0.19
1-year-old released cod	0.49	0.02	0.43	0.06

Data adapted from Jorstad *et al.*, 1994.

caught the following year. Table 4.2 shows the *GPI-1** allele frequencies of the released fish and wild fish of different ages. It is clear that the *GPI-1* 30* allele is rare in all the age classes of wild cod but that it is present at a frequency of 0.49 in the released fish. All released fish (those with the chemical oxytetracycline tag) could be clearly and unambiguously assigned to either the genetically-tagged group of *GPI-1* 30/30* homozygotes or the non-genetically-tagged group. Because equal numbers of wild type and genetically-tagged fish were released, the presence of the *GPI-1* 30* allele at a frequency close to 50% indicates that there has been no significant selection against the genetically-tagged fish – an important fact which needs to be established before such tags can be used commercially.

Genome mapping

A map of the genome is a representation of the position, on each chromosome, of the genes and non-coding DNA markers. Knowing the relative position of a gene of interest makes it easier to find for sequencing and investigating control of its expression, while information about the non-coding elements of the genome will be valuable in assessing the function, if any, of this 97% of the genetic material. The ultimate genetic map is a complete DNA sequence, and this has been completed for *Homo sapiens* in the human genome project. Such detailed knowledge is absent for most other organisms except those that have been used as models for genetic work. Extensive genome mapping has been carried out in two model fish species: the zebrafish (*Danio rerio*) and the medaka (*Oryzias latipes*). Over the last years of the twentieth century the human genome project drove developments in the technology of automated DNA sequencing so fast that we can expect extensive sequencing of the DNA of agriculturally and aquaculturally important organisms over the next 20 years. In the case of commercial fish and shellfish, the main species for which any systematic genome mapping has so far been carried out are the salmon and the Pacific oyster.

How is a genome mapped?

The process of mapping the position of a locus on a chromosome depends on the formation of chiasmata (crossovers) during meiosis. As was noted in Chapter 1, recombination involves the formation of one or more chiasmata along the length of each chromosome pair during prophase of meiosis I. Let us consider the situation of two loci, one of which is situated close to the end of a chromosome arm (locus 1) and the other close to the centromere on the chromosome (locus 2, Fig. 4.1). Both loci are heterozygous with alleles A and B at locus 1 and alleles D and E at locus 2. In one of the pair of chromosomes, alleles A and D are present, and in the other chromosome

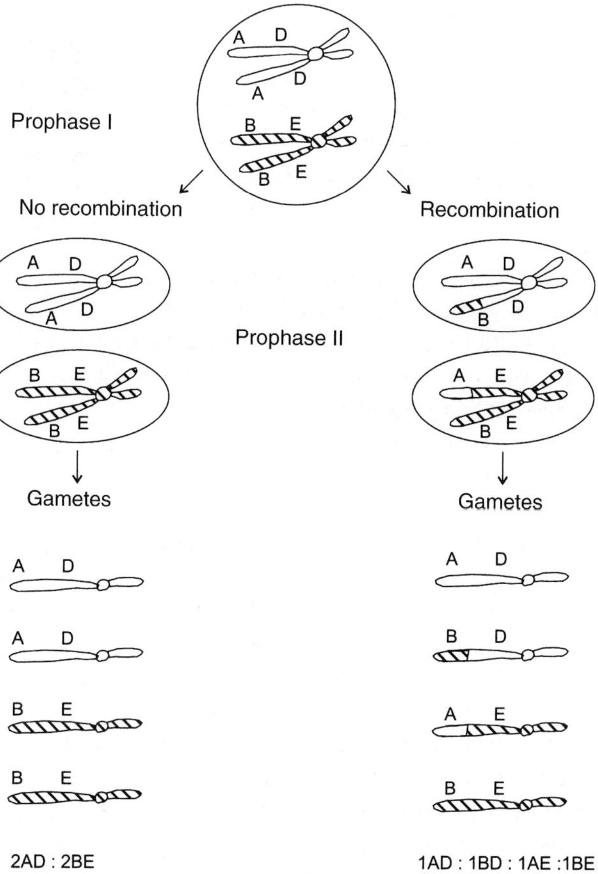

Fig. 4.1 Consequences of recombination during meiosis on the combination of alleles passing to the gametes. A pair of chromosomes are illustrated from an individual which is heterozygous at two loci, one close to the end of the chromosome (alleles A and B) and the other close to the centromere (alleles D and E). Without recombination the gametes contain the two parental combinations (AD and BE) but if recombination occurs there are four gamete types: the two parental types (AD and BE) plus two recombinant types (BD and AE).

of the pair, alleles B and E are present. If no recombination occurs between these two loci during meiosis then there will be 2 A+D gametes and 2 B+E gametes produced. However, if a crossover occurs on the chromosome arm between the positions of the two loci then four gamete types will be produced: A+D, A+E, B+D and B+E. Now consider if we looked at the products of 100 meiotic divisions. If there is never any crossing over between the two loci then there will be 200 A+D gametes and 200 B+E gametes. This tells us something about the position of the loci: they are physically so close together that chiasmata almost never occur between them. In this case the loci are said to be in complete linkage disequilibrium or just complete linkage. Now let us consider if crossovers occurred between the two loci during 60 of the 100 meioses. The result would be 60 each of the recombinant genotypes (A+E and B+D) and 140 each of the undisturbed pairings (the parental genotypes, A+D and B+E). The loci are in partial linkage or linkage disequilibrium. Of course, if two loci occur on different chromosome pairs then they will segregate entirely independently of each other and therefore be in linkage equilibrium.

Now let us make the assumption that recombination is random, by which we mean that there is an equal chance of crossovers occurring at any position along a pair of lined-up chromatids. It follows from this that the number of crossovers will be greater between loci which are far apart, and fewer between loci which are close together. Consequently, the frequency with which loci are decoupled by recombination – their recombination frequency – is directly proportional to the distance between them on the chromosome. If we are able to calculate the recombination frequencies between pairs of loci we can begin to construct a map of their relative positions on the chromosomes (Fig. 4.2). Pairs that always occur in complete linkage equilibrium will be on different chromosomes, while linkage disequilibrium soon makes it clear which loci can be assigned to a particular linkage group. And each linkage group, of course, represents a chromosome pair.

There are complications with this simple outline. First, it transpires that the assumption that recombination events are random along the chromosome arms is not always correct. Particular regions of a chromosome (recombination hotspots) may be more likely to have crossovers than others. Secondly, there is not always only one recombination event along a chromosome arm. There can sometimes be two or more crossovers and when this occurs it can produce parental genotypes between two widely separated loci rather than the expected recombinant genotypes. Nevertheless, despite these limitations, linkage analysis will generally provide the correct information about the order of loci on the chromosomes and give distance estimates of sufficient accuracy to provide a clear framework for further genome analysis.

Genes themselves are frequently monomorphic and therefore not amenable to linkage analysis. In addition, they form only about 3% of the genome and so are too few to act as position markers for sequencing. Instead, much more variable and frequent markers such as microsatellites, or random markers such as AFLPs, are used. The relative positions of cut sites of restriction endonucleases (REs) can also be mapped on chromosomes or mitochondrial DNA, by comparing sizes of fragments produced after incubation of the DNA with different combinations of restriction enzymes. The principle behind this restriction mapping is similar to linkage analysis.

Gene loci scored:

EST-2	esterase
GPI	glucose phosphate isomerase
PGM	phosphoglucomutase
MPI	mannose phosphate isomerase

Recombination frequencies:

Between EST-2 and GPI	= 4.0%
Between EST-2 and MPI	= 35.4%
Between GPI and PGM	= 30.2%
Between PGM and MPI	= 1.2%

Proposed map:

```
MPI  PGM                                        GPI   EST-2

------|--|-----------------------------------------------|---------|--

      0  1.2                                     31.4     35.4
```

Fig. 4.2 An example of the construction of a genetic map from recombination frequencies.

Obviously, REs with very rare recognition sequences have to be used to cut chromosomes into fragments, and restriction mapping has been more commonly used for mitochondrial DNA.

Where we already have some sequence information about a gene or genetic marker its position on the chromosome can be obtained directly by the process of fluorescent *in situ* hybridisation (FISH, Box 4.4).

How do we carry out linkage analysis?

The commonest method involves the analysis of the progeny of experimental crosses carried out between parents of known genotype. Of course this approach cannot be used for analysis of the human genome, where an alternative method is based on family pedigrees. Similarly, in species such as cattle, where there are only a few progeny possible to specific parents, a similar family pedigree method is used. However, no such problems occur with fish and shellfish as there are seldom difficulties in producing large numbers of progeny from pair matings. Indeed, the ability to test hundreds of progeny from a single mating makes the data on linkage disequilibrium far more robust than when only a few progeny can be tested.

As we have seen in the previous section, our assessment of linkage depends upon identifying the di-locus haplotypes of gametes. Although in principle alleles from

Box 4.4 Fluorescent *in situ* hybridisation (FISH)

The FISH technique enables a genetic marker to be precisely located to its position on a chromosome. FISH makes use of the ability of two complementary strands of DNA to form a stable base-paired hybrid molecule.

First, cells are arrested at metaphase of mitosis using colchicine or some other chemical (as in the normal chromosome counting method) and are spread or squashed onto a microscope slide. The DNA in the chromosomes is then denatured into single strands using formamide and a fluorescently labelled probe is added. The probe will be a DNA sequence which is complementary, or almost complementary, to the sequence of DNA being targeted. This could be part of a gene, or another part of the DNA that is non-coding. After washing away any unhybridised probe the chromosome spread is viewed under UV light and the position or positions where the probe has hybridised with the chromosomal DNA can be seen. In this way FISH can be used as part of a gene-mapping programme to physically locate the position of markers used (Fig. B4.4).

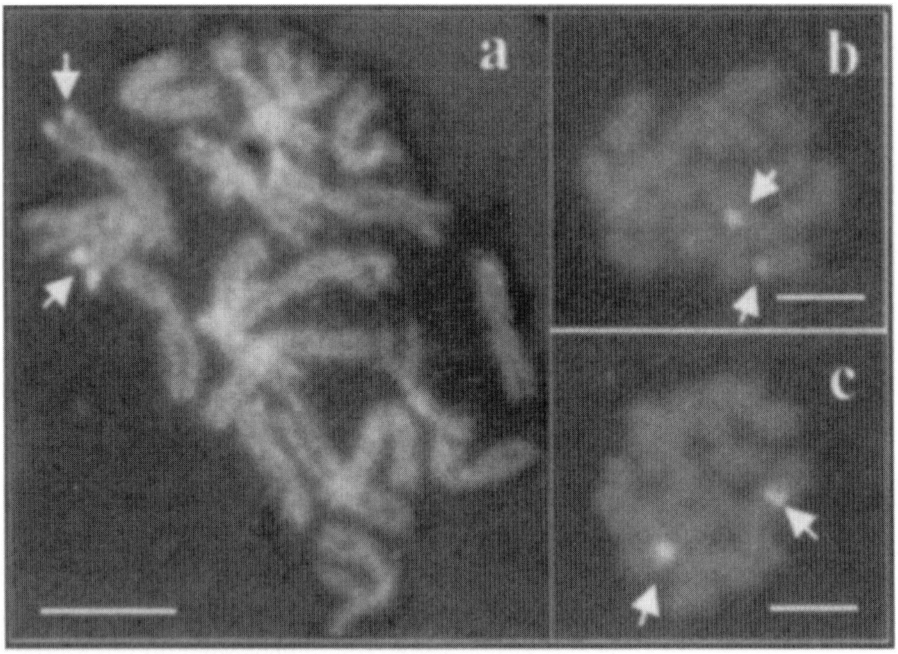

Fig. B4.4 Localisation of the 28S rRNA gene (arrowed) on the chromosomes of the American oyster, *Crassostrea virginica*, using fluourescent *in situ* hybridisation (FISH). Chromosome spreads are from (a) the embryo and (b, c) adult gill cells. (From Zhang, Q., Yu, G., Cooper, R. K. & Tiersch, T. R. (1999) Chromosomal location by fluorescence *in situ* hybridisation of the 28S ribosomal RNA gene of the eastern oyster. *Journal of Shellfish Research*, **18**, 431–435.)

individual spermatozoa could be identified using PCR, this is not feasible in practice, so we do not examine the haploid gametes directly. Instead we identify the di-locus genotypes of the diploid progeny. The complication with this is that the progeny are the result of two meiosis events – one in each parent – and we need to disentangle from the genotypes of the progeny the crossover events that occurred in these meioses. This is most easily done if we are able to select specific parents for the mating, so the standard method is to use a test cross, an example of which is given in Figure 4.3.

(a) **TEST CROSS**

Parent 1 (heterozygous at both loci)	Parent 2 (homozygous at both loci)
AB+DE	**AA+DD**
gametes	gametes
A+D, A+E, B+D, B+E	**A+D**

F1 progeny genotypes

AA+DD	
AA+DE	(recombinant)
AB+DD	(recombinant)
AB+DE	

(b) Example of a **RANDOM CROSS** from which recombination frequencies can be estimated

Parent 1 (heterozygous at both loci)	Parent 2 (heterozygous at one locus and homozygous at the other)
AB+DE	**BC+FF**
gametes	gametes
A+D, A+E, B+D, B+E	**B+F, C+F**

F1 progeny genotypes

AB+DF	
AB+EF	(recombinant)
BB+EF	
BB+DF	(recombinant)
AC+DF	
AC+EF	(recombinant)
BC+EF	
BC+DF	(recombinant)

Fig. 4.3 Identification of recombinant genotypes in (a) a standard test cross and (b) a random cross. Locus 1 has alleles **A**, **B** and **C** and locus 2 has alleles **D**, **E** and **F**. Parent 1 has alleles in the pairs A+D and B+E.

The resulting progeny are genotyped at the loci in question and the proportion of recombinant genotypes calculated. A standard test cross is straightforward to carry out if you already have inbred lines of individuals which are homozygous at most markers and F1 hybrids, produced by crossing such inbred lines, which are heterozygous at most loci. As this is not yet the case for most aquaculture species, random crosses fortunately can also be used to estimate recombination frequencies (Fig. 4.3). The main requirement is that at least one of the parents is doubly heterozygous for the two loci being investigated. Random crosses are much less informative than test crosses because recombinant genotypes will not be identifiable for all loci in the progeny. Nevertheless, during the early stages of domestication of aquaculture species, before inbred homozygous lines and F1 hybrid crosses have been developed, random crossing is the only method available.

The SALMAP project

Mapping the genes in Atlantic salmon, rainbow trout and brown trout has been the focus of a European Community-funded study called the SALMAP project. Currently linkage maps have been constructed for about 300 loci in salmon and rainbow trout and 200 loci in brown trout. The markers used in the maps include a few gene loci, but are mainly microsatellites and therefore not likely to be of direct interest to salmon aquaculture. Nevertheless, some of these neutral loci will act indirectly as genetic markers for genes of interest, including quantitative trait loci (QTL, see Chapter 5).

Current efforts are being directed towards isolating and identifying large numbers of genes in the salmon genome using a strategy based on cloning expressed sequence tags (EST). An expressed sequence is the sequence of a gene with all of the introns removed. Reverse transcription from mRNA using reverse transcriptase produces complementary DNA (cDNA), which is rendered double-stranded using a DNA polymerase to form the EST. ESTs allow protein coding genes to be found from the mRNA pool.

Identification of diseases

One of the major constraints on aquaculture is disease. In the wild there is a natural balance between organisms and their diseases, but this balance is badly upset when organisms are cultured at high densities in enclosed situations. Entire broodstocks or complete production batches can be lost if the early stages of a disease cannot be identified and preventative measures employed. Molecular biology techniques can be used to assist in the early identification of disease-causing organisms such as viruses, bacteria and parasites, as in the following examples:

1. The mycobacteriosis disease of the European sea bass (*Dicentrarchus labrax*), an important Mediterranean culture species, debilitates the fish, slows their growth

and is eventually fatal. Unfortunately, infected fish cannot be identified visually during the early stages of the disease. Following direct sequencing of the host and pathogen 16S rRNA genes, the disease-causing organism was identified as *Mycobacterium marinum*. Primer pairs for the 16SrRNA gene were then used in PCR trials to identify length products specific to the sea bass and products specific to *M. marinum*. A diagnostic 220 bp PCR product from *M. marinum* was identified from a particular primer pair and this method was then developed to reliably amplify the product from blood samples. This technique therefore provides a rapid non-destructive means of monitoring the broodstock of sea bass for the development of mycobacteriosis.

2. A serious viral disease of penaeid shrimp, white spot syndrome virus (WSSV), was first identified in Asia and has recently spread worldwide. The symptoms are white spots inside the carapace and a reddish body discoloration. The disease is very virulent and can wipe out a complete hatchery stock of shrimp within a week or two. Viral DNA has been sequenced and primers have been designed to amplify a short (341 bp) PCR product diagnostic for the disease from shrimp haemolymph. Not only can the presence of the viral DNA be identified, but using quantitative competitive PCR the numbers of viral genomes in an infected shrimp can be estimated. This method uses an internal standard which is co-amplified with the target viral DNA. It is amplified by the same primers but is about 50 bp shorter than the viral PCR product and the amount of this internal standard added to the PCR is known. The two PCR products are separated on agarose gel and the intensity of the stain coloration for each band is measured by a densitometer, allowing a direct estimation of viral numbers based on stain intensity.

Although not providing a cure for the disease, these types of molecular biological tool are valuable additions to the armoury that pathologists need to detect, identify and investigate the progression of diseases in aquaculture organisms.

Further reading

Avise, J.C. (1994) *Molecular Markers, Natural History and Evolution*. Chapman & Hall, London.

Benzie, J.A.H. (ed.) (2002) *Genetics in Aquaculture VII*. Elsevier, Amsterdam.

Doyle, R.W., Herbinger, C.M., Ball, M. & Gall, G.A.E. (eds) (1996) *Genetics in Aquaculture V*. Elsevier, Amsterdam.

Hedgecock, D. & Sly, F. (1990) Genetic drift and effective sizes of hatchery-propagated stocks of the Pacific oyster, *Crassostrea gigas*. *Aquaculture*, **88**, 21–38.

Jørstad, K.E., Naevdal, G., Paulsen, O.E. & Thorkildsen, S. (1994) Release and recapture of genetically tagged cod fry in a Norwegian fjord system. In: *Genetics and Evolution of Aquatic Organisms* (ed. A.R. Beaumont), pp. 519–528. Chapman & Hall, London.

Knibb, W., Colorni, A., Ankaoua, M., Lindell, D., Diamant, A. & Gordin, H. (1993) Detection and amplification of a pathogenic marine mycobacterium from the European sea bass

Dicentrarchus labrax using polymerase chain reaction and direct sequencing of 16SrDNA sequences. *Molecular Marine Biology and Biotechnology*, **2**, 225–232.

McAndrew, B. & Penman, D. (eds) (1999) *Genetics in Aquaculture VI*. Elsevier, Amsterdam.

Ryman, N. & Utter, F. (eds) (1987) *Population Genetics and Fishery Management*. University of Washington Press, Washington.

Tang, K.F.J. & Lightner, D.V. (2000) Quantification of white spot syndrome virus DNA through a competitive polymerase chain reaction. *Aquaculture*, **189**, 11–21.

Chapter 5
Artificial Selection in the Hatchery

So far we have concentrated on markers at the DNA or protein level, how they can be identified and how they can be used to assess genetic variation across the genome. Now we need to consider phenotypic characteristics that are influenced by variation at one or more loci. Examples of such characteristics, or traits, would be colour variants, variations in fin size or shape, body weight at age, disease resistance, numbers of fin rays, age at maturity, larval growth rate and percentage of fat in the meat. There are two main groups of trait – qualitative and quantitative. Qualitative traits can be defined by simple discrete categories (e.g. colour variants) and are often under the control of just one or two loci, while quantitative traits require measurement or enumeration and are usually controlled by many genes (e.g. weight at age, numbers of fin rays).

Qualitative traits

When a breeder discovers an attractive new variant, the first thing to establish is whether the variant is inherited at a single locus in a simple Mendelian way. We have dealt with inheritance of alleles at single loci in previous chapters. However, there is a complication which often arises in the expression of a qualitative phenotypic trait owing to the dominant or recessive nature of the alleles controlling the trait. Let us take a simple example, that of the gold colour variant in carp (*Cyprinus carpio*), which is controlled at a single locus by two alleles, a wild-type allele (W) and a variant gold-producing allele (G). The gold phenotype is only expressed in the fish when the G allele is present in the homozygous condition (i.e. the genotype is GG), while the other two genotypes (WW and WG) will always have wild-type coloration. We describe the W allele as being dominant because its phenotypic expression will occur when allele W is present either in the homozygous or heterozygous condition. Conversely, the G allele is described as recessive because its phenotype is *not* expressed in the heterozygous condition.

Assuming the gold variant is of high market value, how can the breeder produce them in large numbers? Ideally we need to mate a GG male with a GG female to ensure GG offspring; however, in a wild population, there are likely to be only occasional individuals expressing the gold coloration. So, to begin with, what happens if we mate our novel gold-coloured fish to a wild-type fish (Fig. 5.1)? Usually no gold offspring would appear in the F1 generation because the G allele would be rare in the natural population and a wild-type fish taken at random is unlikely to have the G allele. Without the G allele being present in both the father and the mother, no GG genotypes – and therefore no gold phenotypes – will be present among the F1 offspring. However, the F1 offspring will all be heterozygotes with one copy of the

parental genotypes	GG	x	WW	
parental phenotypes	gold	x	wild-type	
F1 genotypes	WG	x	WG	
F1 phenotypes	wild-type	x	wild-type	
F2 genotypes	WW	WG		GG
F2 phenotypes	wild-type	wild-type		gold
Genotype ratio	1 :	2 :		1
Phenotype ratio	3 wild-type		:	1 gold

Fig. 5.1 The phenotypic and genotypic consequences of inheritance of a recessive gold coloration allele (G) and a wild-type allele (W) in carp (*C. carpio*) through the F1 and F2 generations.

G allele and if these are mated together then this will produce GG genotypes (gold phenotypes) among the F2 generation. Assessing the ratio of gold-coloured to wild-type fish in this F2 generation (Fig. 5.1) should allow us to identify the precise relationship between the G allele and the gold phenotype, and GG males and females from this F2 generation can be mated to produce all-gold offspring.

Although this example of gold coloration in carp is straightforward, in the real world of aquaculture things are not usually this simple. Perhaps the colour variant is sex-linked, so that it only occurs in one of the sexes. More commonly, dominance may be incomplete, such that heterozygotes will express a different colour to those expressed in the two homozygote classes. An example of partial dominance is found with the 'gold' phenotype in the tilapia species *Oreochromis mossambicus*, where the wild-type allele (W) is only partially dominant over the gold-type allele (G) – GG genotypes are gold, GW heterozygotes have a 'bronze' skin colour and WW homozygotes express the normal black coloration.

In some situations the two alleles at a locus are co-dominant – that is, both contribute equally to the phenotypic character of the heterozygote. This inheritance pattern is difficult to distinguish from the partial dominance of a single allele, but fortunately the consequences, from the point of view of the breeder, are the same. Obviously, to produce 100% heterozygote phenotype offspring, matings must occur between two homozygotes.

A further level of complexity occurs where two important commercial characters are controlled by two independent unlinked loci (we shall look at interaction between loci later in this chapter). For this example we will consider two phenotypic characters in the guppy (*Lebistes reticulatus*) – gold colour and spinal curvature – controlled independently at separate loci. At the first locus, the dominant wild-type allele (W) produces grey body colouration, while the recessive (G) allele produces gold coloration in homozygous (GG) fish. Because of the complete dominance of the W allele, heterozygous (WG) fish are grey. At the second locus a recessive variant (C) causes curvature of the spine in CC homozygotes. The dominant wild-type allele (Z) gives homozygous (ZZ) and heterozygous (ZC) fish a normal spine. The conventional way to calculate the proportions of offspring genotypes and phenotypes from matings between individuals is to use the Punnett square design where the gamete

Gametes	W Z	W C	G Z	G C
W Z	WW ZZ grey normal	WW ZC grey normal	WG ZZ grey normal	WG ZC grey normal
W C	WW ZC grey normal	WW CC grey curved	WG ZC grey normal	WG CC grey curved
G Z	WG ZZ grey normal	WG ZC grey normal	GG ZZ gold normal	GG ZC gold normal
G C	WG ZC grey normal	WG CC grey curved	GG ZC gold normal	GG CC gold curved

Ratio of different phenotypes:
9 grey normal : 3 grey curved : 3 gold normal : 1 gold curved

Fig. 5.2 Punnett square design showing the consequences of a mating between F1 guppies which are heterozygous at two loci (WG ZC) where the first locus controls gold colour by a recessive allele (G, wild-type =W) and the second locus controls curvature of the spine by a recessive allele (C, wild-type = Z). Phenotypes: grey with normal spine (grey normal), gold with normal spine (gold normal), grey with curved spine (grey curved) and gold with curved spine (gold curved).

types are arrayed as column and row headers and cells within the square are simply completed by the addition of the column and row headings. In the case of guppies, following the strategy of mating between double heterozygotes, the Punnet square illustrates that the offspring phenotypes would be in the ratio of 9 wild type (grey with a normal spine) : 3 gold with a normal spine : 3 grey with a curved spine : 1 gold with a curved spine (Fig. 5.2).

In this example of inheritance of two characters in the guppy, there is no interaction between the two loci. The colour of the fish is not affected by the spinal curvature, and *vice versa*. Where there is interaction between loci in the expression of the phenotypes they control, this is called epistatic interaction. An example of this is the 'pearl' character of Nile tilapia (*O. niloticus*) where the phenotypic opalescent-white character of the scales (designated 'pearl') is controlled epistatically at two loci. Each locus has a wild-type allele and an alternative pearl-type allele and the pearl phenotype is expressed only in individuals that carry pearl-type alleles from both loci. This does not involve the standard type of dominance because pearl-type alleles *can* be expressed in the heterozygous condition. It is simply the presence of the pearl-type alleles at *both* loci that confers the pearl phenotype on its carrier. This is illustrated in a Punnett square (Fig. 5.3) where the wild-type alleles at the two loci are W and Z and the pearl-type alleles are designated P and L respectively.

Gametes	W Z	W L	P Z	P L
W Z	WW ZZ wild-type	WW ZL wild-type	WP ZZ wild-type	WP ZL pearl
W L	WW ZL wild-type	WW LL wild-type	WP ZL pearl	WP LL pearl
P Z	WP ZZ wild-type	WP ZL pearl	PP ZZ wild-type	PP ZL pearl
P L	WP ZL pearl	WP LL pearl	PP ZL pearl	PP LL pearl

Ratio of phenotypes : 7 wild-type : 9 pearl

Fig. 5.3 Punnett square illustrating inheritance of the 'pearl' phenotype in the Nile tilapia (*O. niloticus*) based on variation at two epistatic loci. At the first locus the wild-type allele is designated W and the pearl-type allele is indicated by P. At the second locus the wild-type allele is designated Z and the pearl-type allele is given as L. Matings were between doubly heterozygous F1 individuals (WP ZL × WP ZL).

A further example of epistatic interaction involves the pattern of the scales in the common carp (*C. carpio*), one of the world's most important cultured fish. Carp with a reduced scale pattern are highly valued in the European market while Asian consumers prefer the wild-type scale pattern. At one locus (wild-type allele = W, variant allele = S) the S allele reduces the overall scaliness of the fish such that SS homozygotes exhibit a phenotype called 'mirror'. At the other locus (wild-type allele = Z, variant allele = N) the presence of the Z allele modifies the pattern of scales in heterozygotes, but is lethal in the homozygous state (ZZ). This locus is epistatic to the W/S locus – it also modifies scale pattern – and it shows incomplete dominance as although ZZ homozygotes die, heterozygotes (ZN) survive and NN homozygotes have wild-type scale pattern. Figure 5.4 illustrates the ratios of the different phenotypes produced by mating carp which are double heterozygotes at these loci. The phenotypes are 'normal scaled'; 'line', where the scales are mostly restricted to the lateral line; 'mirror', where a few scales are scattered across the body; and 'leather', where scales are virtually absent.

So far we have assumed that scaling pattern loci in carp only affect scale density and pattern. However, allelic variation at a single locus may have effects on more than one biochemical pathway and this can have subtle effects on other phenotypic characters. This is called pleiotropy and there are a number of performance-based pleiotropic effects on mirror, line and leather carp. The wild-type scaled carp generally grow, survive and resist disease better than do mirror, line or leather carp.

Gametes	W Z	W N	S Z	S N
W Z	WW ZZ lethal	WW ZN line	WS ZZ lethal	WS ZN line
W N	WW ZN line	WW NN wild-type	WS ZN line	WS NN wild-type
S Z	WS ZZ lethal	WS ZN line	SS ZZ lethal	SS NZ leather
S N	WS ZN line	WS NN wild-type	SS ZN leather	SS NN mirror

Phenotypes ratio : 6 line : 4 lethal : 3 wild-type : 2 leather : 1 mirror

Fig. 5.4 Punnett square illustrating inheritance of wild-type, mirror, line and leather phenotypes of carp (*C. carpio*). The mating is between fish which are double heterozygotes at two loci. One locus (wild-type allele = W, variant allele = S) controls scale density and the other, an epistatic locus (wild-type allele = Z, variant allele = N), controls the pattern of scales. Allele Z is a partially dominant lethal allele.

Globally, the development of colourful phenotypes of fish is of great economic importance. The Taiwan Fisheries Research Institute has spent many years developing the potential of red variants of *O. mossambicus*, that are controlled by recessive alleles. They developed hybrids with *O. niloticus* and, following several generations of artificial selection, produced stable red and white strains of the hybrids.

There are as yet few examples of the commercial use of qualitative traits in shellfish, such as the golden coloured Pacific oyster (*Crassostrea gigas*) from Tasmania, or the 'notata' form of the hard-shell clam (*Mercenaria mercenaria*) in the USA.

Further descriptions of the profusion of albino types and colour variants in tilapias and other fish such as koi carp are outside of the scope of this volume, but all are based on the principles of Mendelian inheritance, albeit complicated by partial dominance, the involvement of more than one locus, more than two alleles per locus, epistasis, sex linkage or pleiotropy.

Quantitative traits

Now we move on to the consideration of traits that are usually controlled by more than just one or two genes – polygenic traits. The study of these polygenic, or quantitative, traits is the basis of quantitative genetics. Unlike qualitative trait loci, genes that are assumed to be influencing or controlling a quantitative trait are very rarely identified. They are given the overarching title of quantitative trait loci (QTL).

Of course, although a quantitative trait will be influenced through its QTL it is also very likely that the trait will be influenced by the environment in which the organisms exist. A well-fed prawn will grow faster than a poorly-fed prawn, regardless of its genotype. When we use the term 'environment' here, this covers all aspects of the way in which an organism has interacted with the physical and biological environment around it during its life up to the time that the trait is measured or counted. It will also include the internal cellular environment in which the proteins produced by the genes have to operate. So when considering the influence of the environment on a particular trait it is not enough to consider only the tanks in which fish might be swimming around or simply to count the number of mysids being fed to a growing lobster larva. Environmental interactions are extremely complex and many aspects of such interactions will be invisible, undetectable, unexpected or simply unknown.

Biologists study quantitative trait variation for a number of reasons, but we shall concentrate here on the use of quantitative genetics as a tool for understanding and implementing selective breeding of aquatic organisms in commercial culture. Selective breeding of some form or other has been undertaken for hundreds, sometimes thousands, of years in agricultural and domestic plants and animals. The results are all around us and we take them for granted. Cattle have been bred either for meat or for dairy use; sheep have been bred for wool or meat and for mountain or lowland. There are hundreds of varieties of cereals and crops that have been selectively bred for particular soils, seasons or production characteristics. Perhaps the most striking range of phenotypic variation obtained by artificial selection within a single species is evident in the domestic dog (*Canis domesticus*) ranging from the tiny Chihuahua to the giant Irish Wolfhound.

Among aquatic organisms, probably the first to be intensively bred and managed were the carps (*Cyprinus* spp.) and in the Far East intensive selection has taken place over hundreds of years for unusual colour variants leading to a highly valuable ornamental fish market. However, very few aquatic species brought into commercial culture have been subjected to the application of scientifically-based artificial selection, particularly quantitative genetics. Indeed, it is only really within the past 25 years that any studies of quantitative traits in aquatic species have been carried out, and these have been limited to a few species of fish and shellfish. In this chapter we will illustrate and discuss how quantitative genetics and the investigation of QTL is likely to be used in the process of artificial selection and the development of commercially-adapted strains of aquatic species as part of their domestication.

What kinds of traits are important?

For the evolutionary biologist all phenotypic traits are important as they are almost all likely to be relevant to the genetic fitness of the organism. For the aquaculturist, particular traits that are relevant to production will be of interest. Such traits include growth rate, body weight at marketing, feed conversion efficiency, disease resistance, flesh quality and age at maturation.

Variance of a trait

In a sample of organisms, measurements of a continuously variable trait such as weight at age will tend to be distributed normally (Fig. 5.5) and the distribution can be described using standard statistical notation. There will be a mean (symbol = $\overline{x}$), a standard deviation from the mean (symbol = s) and a variance (symbol = s^2) which is the square of the standard deviation from the mean. The variance is simply a number which represents the spread of measurements to either side of the mean: the greater the spread of measurements, the larger the variance. Some part of that variance will be the result of genetic differences between individuals and the remaining part will be the result of differences in the environments experienced by different individuals. We need to know the relative influences of these two factors – there is no selective advantage in breeding from the largest members of a broodstock if they are only the largest because of the environment in which they grew up.

The total variance of a trait can be divided into that produced by genes and that produced by the environment such that:

$$V_p = V_g + V_e + (V_g \times V_e)$$

where the total phenotypic variance is V_p, the variance produced by genetic differences among individuals is V_g and the component of variance produced by environmental differences is V_e. Some part of the variance will be produced by gene–environment interaction, so the term $(V_g \times V_e)$ is included in the formula. Because this component is very difficult to identify, and is thought to be usually quite small, for all practical purposes it is assumed to be negligible. There is one more element of the variance that should be mentioned. It is the variance that is the result of

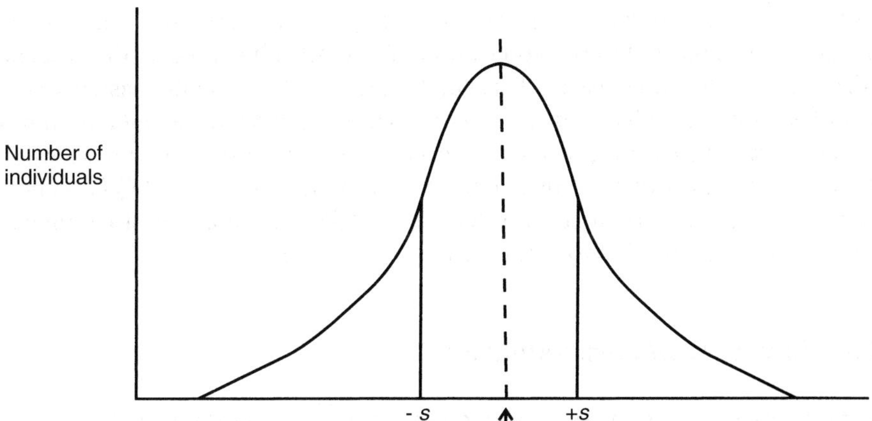

Fig. 5.5 Normal distribution of a trait in a sample of organisms, showing the mean and standard deviation (s). The variance of the trait is equal to the square of the standard deviation (s^2).

maternal factors, both genetic and environmental (e.g. mitochondrial DNA, egg size, brooding behaviour, nesting location), and that can represent a considerable fraction of V_g and V_e.

The proportion of the variance of a trait which is under genetic control is termed the heritability, denoted by the symbol h^2. The broad-sense heritability is:

$$h^2 = V_g / V_p$$

which represents the proportion of the total variance (V_p) which is produced by genetic variance (V_g).

The variation produced by genes may come from three different sources. First, the variation may be related to the straightforward presence or absence of particular alleles at the QTL. This is the additive genetic variance (V_a) and is the most important element of genetic variation for the aquaculture breeder as the presence or absence of particular alleles is a character which is passed unchanged to the next generation.

A second cause of genetic variation is due to the presence or absence of particular genotypes at QTL. For example, a particular heterozygous combination of alleles at a locus may confer an advantage or disadvantage on an individual with respect to a particular trait. This is dominance genetic variance (V_d) and is much less amenable to simple artificial selection because genotypes are broken down during meiosis and put together again in different combinations in the next generation. Therefore, breeding from an individual which, because it is heterozygous at one or more QTL, has a high ranking for a trait, does not guarantee inheritance of the advantageous heterozygosity in its offspring. However, crossing between inbred, highly homozygous lines does guarantee predictable heterozygosity in the offspring and such 'F1 hybrids' are commonly used in plant and animal breeding. As yet, development of such inbred lines in aquatic species is in its infancy but there is great potential for producing inbred lines in such species using ploidy manipulation techniques (see Chapter 6).

The third component of genetic variation is produced by interactions between loci and is called epistatic or non-allelic interaction genetic variance (V_i). Here we are considering that an individual may be highly ranked for a trait as it possesses particular combinations of genotypes across two or more QTL. For example, it may be an AC heterozygote at one locus, an EE homozygote at a second and an PQ heterozygote at a third, and it is this particular combination that gives it the high ranking. It is easy to see that only a very complicated controlled breeding scheme designed for such individuals could guarantee this particular combination in any of the offspring.

So the genetic component of variance can be partitioned such that:

$$V_g = V_a + V_d + V_i$$

and the ratio of additive genetic variance (V_a) to the total phenotypic variance is called the narrow-sense heritability:

$$h^2 = V_a / V_p$$

Values of h^2 (either broad-sense or narrow-sense) theoretically vary from zero, where phenotypic variance is entirely the result of environmental effects, to 1.0, where all the variance is the result of genetic effects. The higher the heritability, the greater will be the resemblance between parents and offspring. It is this resemblance between related individuals that is used experimentally to estimate heritabilities of important traits.

How can we estimate narrow-sense heritability?

A number of methods can be used to estimate h^2, such as regression of full-sib means on the mid-parent mean or comparison of full-sibs and half-sibs using analysis of variance (Box 5.1). Given that much of quantitative genetics was developed around agricultural animals, the reader will find that in the literature male parents are called sires and female parents are called dams. Self-fertilisation is included in some of the experimental designs that have been used to estimate heritabilities in plants (in aquaculture these methods could be applied to simultaneous hermaphrodites such as scallops, or, using cryopreservation of gametes (Box 5.2), to sequential hermaphrodites such as oysters). Homozygous inbred lines are employed in other methods for the estimation of heritability and, as indicated elsewhere, development of such inbred lines is one aspect of the domestication of aquatic organisms which is likely to take place in the near future.

In order to produce reliable estimates of h^2 using analysis of variance of full-sib and half-sib hatchery crosses, all crosses must be performed at the same time and all offspring from all families must be reared in the same environment. Achieving simultaneous crosses is relatively simple in many fish and molluscs, but can be a real problem in other species. Some bivalves, such as scallops, are simultaneous hermaphrodites and great care is needed to avoid self-fertilisation in hatchery conditions. Female and male flat oysters are indistinguishable by appearance and the females brood the larvae in their mantle cavity. Many pairs of oysters therefore need to be set up to ensure that a sufficient numbers of male and female pairs. Also, the fact that individual oysters will change sex during their life is not helpful! Strip spawning of catfish males is not usually possible, so the males have to be allowed to perform multiple matings separated by varying periods of time. This results in uncontrollable differences in size or age between half-sib families and may create error in the calculation of h^2. Females of some prawn species cannot normally be induced to ovulate at a specified time and males cannot be strip spawned. In fact they deposit their spermatophore on the cephalothorax of the female just following her prespawning molt which makes control of timing of crosses very difficult. In some crustaceans, such as the freshwater crayfish, the females mate at intervals with several different males and retain their attached spermatophores until they ovulate. Consequently, not only is it very difficult to organise the timing of the crosses, but also the identity of the male parent needs to be established.

Once families have been produced, they must be kept in replicated containers in the same environment. Environmental elements such as water quality, food quality and quantity, temperature, stocking density and level of disturbance must be monitored and held constant. The potential for variation of any of these factors generally means that statistical analysis of large numbers of families is required to provide

Box 5.1 Estimation of narrow-sense heritability

Parent–offspring regression

If the variance in a trait, for example weight, is mainly of genetic origin then we would expect heavy parents to produce heavy offspring. If weight is mainly environmentally controlled, then this would not be the case. This is the principle behind parent–offspring regression, where the mean of a trait in offspring is regressed against the mid-parent value for the trait. A series of matings are carried out with the offspring from all families being reared under the same environmental conditions. Each cross produces one point on the regression. The slope of the regression is equal to the heritability (h^2) of the trait. Figure B5.1 shows an example of weight-at-age data and the regression line has a slope of $0.33 = h^2$.

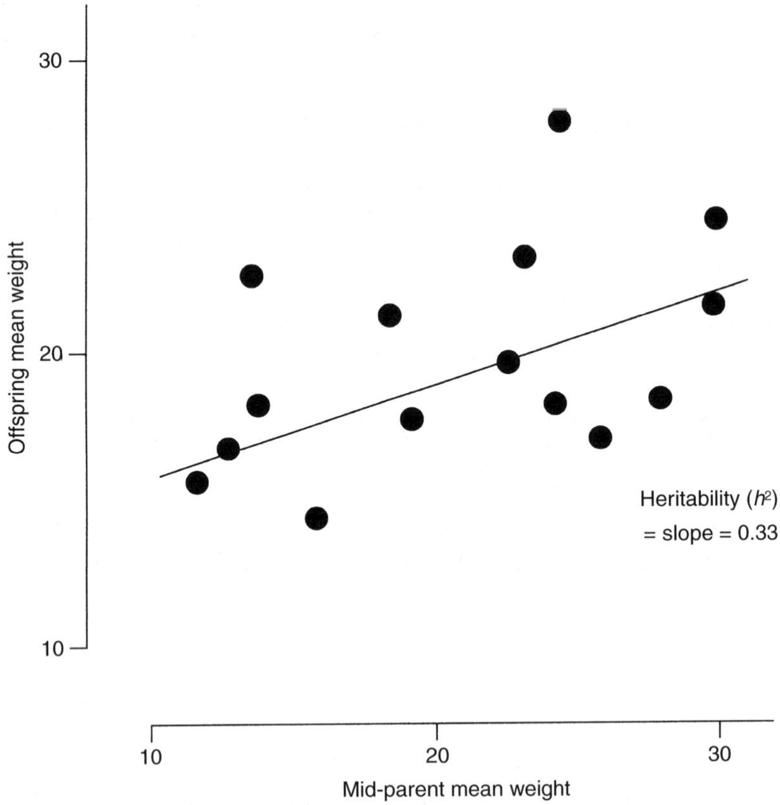

Fig. B5.1 The use of parent–offspring regression to estimate the narrow-sense heritability (h^2) of a trait.

Full-sib and half-sib mating designs

As with parent–offspring regression, it is the resemblance between relatives that provides a means of estimating the heritability of a trait. If the trait is under strong genetic control (high h^2) the variation in that trait within one family is going to be less than the variation between unrelated individuals. Some of the experimental designs for estimating narrow-sense heritability are called North Carolina experiments (NCI, NCII NCIII) because they were first described by scientists from North Carolina in the 1950s. Here we will consider the NCI method where a number (n) of males (sires) are each mated to two or more (f) females (dams) with no dam being mated to more than one sire. This will produce ($n \times f$) full-sib families. In order to arrive at estimates of h^2 with reasonably low standard errors, the number of families needs to be as high as possible. Each set of families with the same sire constitutes a half-sib family group. So we have n half-sib groups, each made up of f full-sib families. Measurements of the trait are made from r samples taken from each family and there are therefore $n \times f \times r$ progeny to be scored and analysed. The data are analysed by hierarchical or nested analysis of variance (ANOVA). The generalised results of such an ANOVA are given in Table B5.1.

Table B5.1 General ANOVA of a North Carolina I experiment to estimate heritability of a trait.

Source	d.f.	MS	Estimates of variance components
Between sire HS family groups	$n-1$	MS_s	$\sigma^2_w + r\sigma^2_{s/d} + rf\sigma^2_s$ $(=\sigma^2{}_{total})$
Between FS families within sire HS family groups	$n(f-1)$	$MS_{s/d}$	$\sigma^2_w + r\sigma^2_{s/d}$
Within FS families	$nf(r-1)$	MS_w	σ^2_w
Total	$nfr-1$		

HS = half-sib, FS = full-sib, d.f. = degrees of freedom, MS = mean square, σ^2 = variance, s = sire component, d = dam component, w = within FS families.

The three components of variation, offspring, dam and sire (σ^2_w , $\sigma^2_{s/d}$ and σ^2_s), can thus all be calculated and used to estimate the additive genetic variance (V_a), since:

$$V_a = 4\sigma^2_s$$

and, since $h^2 = V_a/V_p$, narrow-sense heritability is given by:

$$h^2 = 4\sigma^2_s/\sigma^2_{total}$$

For full details of the various designs for estimating heritabilities the reader is referred to Kearsey & Pooni (1996) or Falconer (1989).

Box 5.2 Cryopreservation

Cryopreservation was first used to assist in reproductive strategies for agricultural animals and the basic methods have since been transferred to aquaculture. This technology enables the aquaculturist to produce seed stock out of season, but also, importantly, allows for the long-term storage of valuable genetic material. Such material has value in selective breeding plans, in the production of hybrids and as a conservation tool for endangered or rare species.

Care must be taken when freezing live material since the water it contains expands during ice crystal formation, potentially rupturing the cell – which is why freezing strawberries makes them go mushy (peas can be successfully frozen and thawed because, as seeds, they have a much lower water content). Cryopreservation thus requires a careful management of cell water content, through the use of cryoprotectants. Essentially gametes or embryos are bathed in an 'anti-freeze' solution containing cryoprotectants and are then frozen down to the temperature of liquid nitrogen (–196°C). When required for use, the material is gently thawed.

Cryoprotectants can be of two types, penetrative or non-penetrative, depending upon whether they pass easily across the cell membrane or not. Dimethyl sulphoxide (DMSO), glycerol and ethylene glycol are examples of penetrative chemicals while larger molecular-weight molecules such as carbohydrates and proteins generally do not penetrate the cell membrane over short periods.

As the cells cool below 0°C, water begins to freeze out of solution and this increases the concentration of the remaining solution. With further reduction in temperature the solution gets more concentrated as more water freezes out. Eventually a point is reached – the eutectic point – at which the whole solution freezes. The cryoprotectant mixture used must effectively protect the cells against damage during this cooling process. Gametes and other cells have an internal concentration of solutes (an osmolality) that must be balanced against a similar concentration of solutes outside the cell.

Initially, when cells are placed in a cryoprotectant, the cells shrink as water leaves the cell to dilute the cryoprotectant on the outside and then the cryoprotectant enters the cell until equilibrium is reached. To reduce the mechanical dangers of this, cryoprotectant can be added at slowly increasing concentrations. If cells are cooled too rapidly, lethal ice crystals can form, but cooling too slowly allows extended contact with the cryoprotectant solution that can be toxic. Once the cells have reached their eutectic point and no more water comes out of solution (–35°C to –80°C), they can be rapidly cooled to –196°C by plunging them into liquid nitrogen.

Non-penetrative cryoprotectants are often used to offset problems of cell expansion and the formation of ice crystals during thawing and, as with cooling, the rate of temperature change can be critical.

The sperm or spermatophores of a range of fish and shellfish can be cryopreserved and an example is given here for the cryopreservation of trout spermatozoa.

Semen is stripped from the trout by abdominal massage and then diluted with a cryoprotectant solution containing DMSO and glycerol. The diluted semen is drawn up into narrow 'straws' and these straws are then frozen in the vapour of liquid nitrogen in an insulated box. The freezing process takes about 3 to 4 minutes before plunging into liquid nitrogen. Thawing is accomplished in about 30 seconds by holding the straws at 25°C and the rate of thawing is critical to obtaining good viability of spermatozoa.

Attempts to cryopreserve fish eggs and embryos have not been very successful, probably due to the size of the eggs and the presence in many species of a relatively impermeable egg membrane. However, in recent years methods for the cryopreservation of the eggs, embryos and larvae of many invertebrates such as oysters, mussels, rotifers and ragworms have been published.

meaningful h^2 estimates. This requires a considerable investment in space and facilities which is often not feasible in a small-scale hatchery.

On the other hand, if individuals could be tagged or individually identified in some way then they could all be reared together in a common environment which removes the need for extensive hatchery space and greatly reduces the statistical problems caused by variable environments. But how can you tag individual fish or shellfish at a small size? When fish reach a certain size they can be heat- or cold-branded or their fins can be clipped, but, of course, trying to mark the carapace of crustaceans for identification purposes is useless as they moult regularly. One method used for identifying individual fish or crustacea is that of inserting a passive integrated transponder (PIT) tag into the flesh, but this would be very expensive for marking large numbers of juveniles in a commercial situation, and is really only effective for marking adult broodstock.

The development of highly variable genetic markers that can be detected from minute quantities of tissue provides a neat solution to this problem. In particular, microsatellites (Chapter 2, Box 2.7) are used regularly to identify young salmon into their families in Canadian, Norwegian and Scottish salmon hatcheries. A suite of just a few highly variable microsatellite loci is usually sufficient to enable more than 95% of individuals to be identified to their family in a 10 × 10 cross (100 families). At a highly variable microsatellite locus individual parents are likely to differ in at least one of their alleles from all other parents. When this is extended to several highly variable loci the multi-locus genotypic signal of each parent is almost certain to be distinct, and therefore the multi-locus genotypic signature of the offspring of crosses between these parents can be identified.

Although this technology is currently beyond the scope of all but the largest hatcheries there are now a number of genetic technology companies which can provide hatcheries with precise identification of their young fish or shellfish at reasonable cost. Over the coming decades further refinement of genetic methods will tend to lower rather than raise the cost of this type of activity and possibly bring it within the range of many commercial hatcheries.

Table 5.1 Estimates of heritability ($h^2 \pm$ standard error) for characters of importance in salmonid aquaculture

Trait	Atlantic salmon	Rainbow trout
Body weight at harvest	0.35 ± 0.10	0.21
Age at maturation	0.15	0.05
Fat percentage	0.30 ± 0.09	0.47
Flesh colour	0.09 ± 0.05	0.27
Survival overall	0.00 ± 0.02	0.16 ± 0.03
Furunculosis survival	0.04 ± 0.17	–

Data from Gjedrem, 2000.

Table 5.1 illustrates heritabilities in Atlantic salmon and rainbow trout for certain traits of importance in salmonid aquaculture. Apart from the observation that none of the heritabilities are greater than 0.5, and most are much less than this, two factors stand out in these figures. First, heritabilities can be very different for the different traits. In Atlantic salmon, approximately 35% of the variation in growth rate is controlled by additive genetic variation ($h^2 = 0.35$), but none of the variation in overall survival is so controlled ($h^2 = 0.0$). Secondly, even though these two species belong to the same genus, there are big differences between them in the heritability of certain traits. For example, the age at which Atlantic salmon mature appears to be more strongly under the control of additive genes than the age at which rainbow trout mature. In contrast, flesh colour has a much higher heritability in rainbow trout than in Atlantic salmon.

These observations reinforce the message that heritabilities are not easily predictable: indeed, differences in heritabilities between different populations within a species are commonly observed. Also, as might be expected, heritability estimates for a single population or broodstock will vary if trials are conducted in different environments.

Variation in heritability means that some traits are far more amenable to rapid change by selection than others – the relatively high h^2 of salmon body weight at market size ($h^2 = 0.35$) implies that this trait will respond well to selection. This has been confirmed by the genetic gains in growth rate ranging from 11 to 15% per generation achieved in the National Norwegian Breeding Programme for salmon which was initiated in the 1970s.

Correlated traits

The phenotypes of organisms comprise thousands of characters and many of these characters are interlinked in one way or another. Therefore, there is a danger that if you select for one particular trait which is of value in aquaculture, you might inadvertently be selecting for another trait which could be deleterious. For example, fast growth rate in salmon could be correlated with age at maturity such that, beyond a certain point, the advantages of fast growth will be outweighed by the tendency of these fish to mature earlier than wanted by the fish farmer. Fortunately, where such tests have been carried out in fish, high correlations have been found between feed conversion efficiency and growth rate and low correlations between growth rate, and survival. One unwanted correlation, however, is between body weight and fat content in salmonids which means that it is not easy to artificially select for increased growth rate without also increasing the body fat content.

In channel catfish (*Ictalurus punctatus*) the time of maturation is dependent on age rather than size. This can create difficulties as selected fast growing fish in some strains reach such a large size before becoming mature that they are too big to be accommodated in conventional spawning refuge chambers. However, one useful genetic correlation in this species is that between the traits of body density and dress-out percentage (the proportion of the fish's weight remaining after gutting). Normally, dressout percentage can only be assessed after slaughter but, because of the strong genetic correlation of this trait with body density, measurements of body density on living fish can be used to estimate dressout percentage.

What types of artificial selection are there?

Mass selection – also confusingly called individual selection – is the simplest type of artificial selection and involves selecting the best-performing individuals for the trait in question from a population, and breeding from them. It usually results in an increase in the mean of the trait in the offspring compared with the parental population. Mass selection can be performed for several generations until the desired shift in the mean value of the trait has been obtained.

The selection differential, s, is the difference between the mean value for the trait in the parental population and the mean value of the individuals used for breeding. If the heritability of the trait is known, then the expected improvement in the mean value of the trait (the response to selection, r) can be estimated as the product of the heritability and the selection differential (Box 5.3). In addition, if an estimated heritability is not known, a 'realised' heritability can be calculated following an artificial selection trial where the selection differential is known and the response to selection can be measured in the offspring.

One of the difficulties associated with predicting the response to selection is that h^2 will usually have been estimated under experimental conditions that may have been very different from the production-scale environment under which the selection is carried out. Because heritabilities may vary considerably between environments the predicted response may not be achieved. In addition, the h^2 value estimated for one population may not apply to another because the amounts of additive genetic variance for the trait may be very different between the two.

Mass selection is simple to carry out and easy to manage in a hatchery situation, but it is only really effective when the true value of h^2 is greater than around 0.3 and when large selection differentials can be employed. Large selection differentials

Box 5.3 Response to selection and realised heritability

The most accurate way to determine the heritability of a trait is to actually carry out selection trials and measure the response to selection for the trait. Figure B5.3 illustrates an example where we have a population of organisms with a mean weight (at 24 months old) of 20 g and only individuals over 21.5 g in weight are selected for spawning to produce the next generation. The difference between the mean weight of all the original population and the mean weight of those selected (22.1 g) is called the selection differential (s = 2.1 g). When the offspring are 24 months old they are weighed and the difference between the mean weight of the offspring (21.0 g) and that of the original population (20.0 g) is the response to selection (r = 1.0 g). When we know the selection differential (s) and the response to selection (r) the realised heritability can be calculated as r/s. It follows also that when we already have an estimate of h^2, and can choose a selection differential, then we can estimate an expected response to selection.

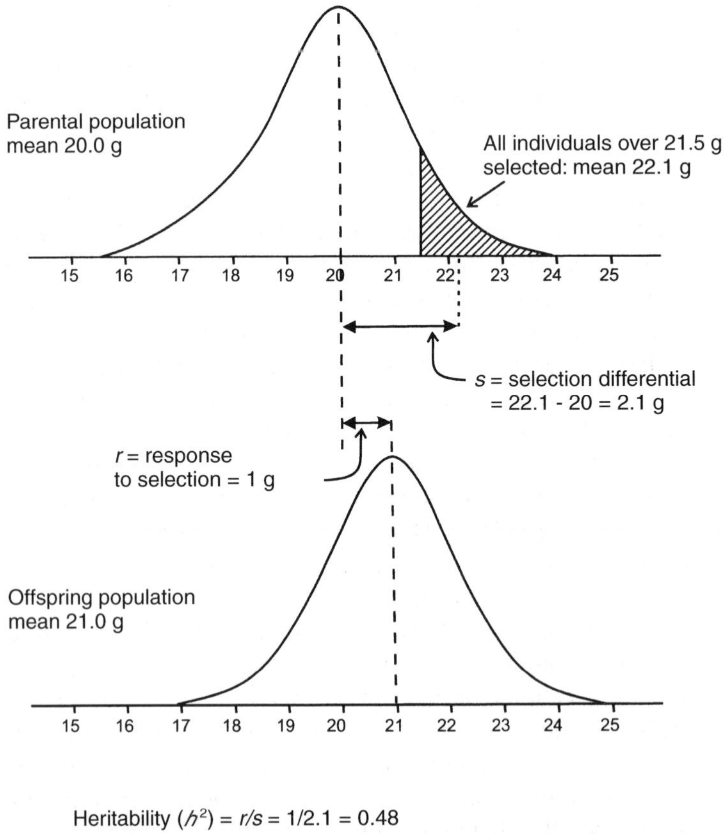

Fig. B5.3 Response to mass selection.

There is, however, a potential problem with this method. The environmental conditions during the growth of the offspring may not be the same as those in which the original population were grown. In such a case, any apparent response to selection might simply be a response to improved rearing conditions and have little to do with additive genetic variance. To obviate this problem it is critical to compare response to selection in the selected population against a control population derived from parents taken at random from the original population. Therefore, whenever a group of selected parents are spawned, a control group of a similar number of randomly chosen parents should also be spawned and their offspring reared in identical environments.

cannot be repeatedly employed when there are only small broodstock populations available because only a few individuals would be mated and inbreeding would result.

When the h^2 value for a trait is low (<0.30), improved selection response can be achieved using family selection. This involves choosing entire families, usually groups of full-sibs or half-sibs, rather than individuals for breeding. The families are selected on the basis of the mean trait value across the whole family. An individual's trait value is not considered in the selection process: the individual just has to be from a particular family. An important limitation of this method is the need to maintain a number of different families separately as broodstocks. Strategies to reduce inbreeding will also be required. Nevertheless, when h^2 is low, family selection produces a greater response than mass selection.

An alternative approach that can be usefully employed is within family selection where individuals are selected for breeding on the basis that they are highly ranked for the trait in question within their family. In this situation, the family mean for the trait is not considered. Although individuals from poorly-performing families may have a low value for the trait relative to the population as a whole, their inclusion in the breeding scheme provides an effective counterbalance to the inbreeding that can occur when family selection alone is employed. Because of this, fewer families need to be maintained and this is a real advantage in most aquaculture enterprises where the cost of holding broodstock families separately can be prohibitive.

A more sophisticated method of selecting individuals for breeding is the use of marker assisted selection (MAS). During the process of gene-mapping the genetic markers used can be tested to ascertain whether possession of particular alleles at these loci is correlated with performance. Any markers that are found to show such an association (QTL) can be of value in artificial selection. MAS simply involves scoring broodstock individuals and selecting them on the basis of their genotypes at QTL.

In any breeding programme, individuals are going to be of different value depending on how much improvement in a trait their offspring produce. This is an individual's breeding value. More than one trait of importance to aquaculture production can be used to estimate the breeding value of an individual. An individual can therefore be given an index that is the weighted sum of its breeding value for all of the traits where performance of its offspring is known. A further refinement is to weight these breeding values on the basis of the relative importance of the traits from the aquaculture point of view. Thus selection based on a carefully weighted index of the breeding value of individuals is a desirable endpoint in any breeding programme. However, considerable investment has to be made before such index-based selection is possible and it is unlikely that many aquaculture enterprises will be able to reach that endpoint. In particular, the short lives of individual fish or shellfish represent a practical limitation to development of this level of sophistication, although this could be offset by the cryopreservation of gametes (Box 5.2).

Multiple traits can be dealt with in other ways besides using breeding values. Tandem selection involves selecting for one trait in one generation and another trait in the next. Independent culling requires that each individual selected for breeding has to score well for more than one trait. Finally, if two traits are known to be highly genetically correlated then it is possible to select for one trait (the secondary trait) as a means of improving the other (the primary trait); such indirect selection is particularly useful when heritability for the primary trait is low but that for the secondary trait is high.

What about realised heritabilities?

Although we can estimate heritabilities by the various methods described in Box 5.1, the only way to establish the true, or realised, heritability of a trait is to carry out selection trials and measure the actual response (Box 5.3). Remember, however, that the realised heritability is only true for the particular environment in which the selection was carried out and for the particular strain or broodstock that has been selected. There are many instances where researchers in different laboratories have published very different realised heritabilities for the same species in apparently similar environments. Some realised heritabilities for commercially important traits in fish, molluscs and crustaceans are given in Table 5.2.

Setting up a breeding programme

The initial requirement for any breeding programme is to acquire certain basic information about the population of fish or shellfish involved in the programme. First, of course, there must be a recognised method of culture that closes the life cycle. Next, there should be variation associated with the traits of commercial interest, and the extent of the phenotypic and genetic variation of these traits must be quantified. This is done by establishing, for each important trait, its range, mean, variance and heritability. When this background knowledge is in place then the breeding goals can be more precisely defined. In principle, all economically valuable traits that can be clearly measured should be included in the breeding goal. Nevertheless, breeding goals must be realistic, practicable and take account of costs, space available, generation time and predicted gain. The planning of a breeding programme is a major task and requires meticulous care in designing every aspect. For example, broodstock

Table 5.2 Some examples of realised heritability (h^2) for quantitative traits in fish, molluscs and crustaceans

Group and species	Trait	h^2
Fish		
Carp (*Cyprinus carpio*)	Body shape	0.42 ± 0.03
	Antibody production	0.37 ± 0.36
Tilapia (*Oreochromis niloticus*)	Growth rate	0.12
Golden shiner (*Notemigornus crysoleucas*)	Growth in length	0.42
Atlantic salmon (*Salmo salar*)	Growth	0.2 to 0.45
Molluscs		
Bay scallop (*Argopecten irradians*)	Growth	0.21
Hard clam (*Mercenaria mercenaria*)	Growth at medium density	0.40
	Growth at high density	0.0
Chilean oyster (*Tiostrea lutaria*)	Live weight at 40 months	0.43 to 0.69
European flat oyster (*Ostrea edulis*)	Growth in weight and length	0.11 to 0.24
Oyster (*Saccostrea cucullata*)	Growth to market size	0.28 ± 0.01
Crustacea		
Marron crayfish (*Cherax tenuimanus*)	Growth and tail size	0.30 to 0.60
Redclaw crayfish (*Cherax quadricarinatum*)	Growth	0.24 ± 0.06
Pacific white shrimp (*Litopenaeus vannamei*)	Weight gain	0.40 ± 0.06
	TSV* resistance	0.09 ± 0.03
Prawn (*Macrobrachium nipponense*)	Larval tolerance of fresh water	0.24 ± 0.07
Prawn (*Macrobrachium rosenbergii*)	Cold tolerance	0.50 to 0.90

* TSV = Taura syndrome virus.

individuals need to be made identifiable and information about their breeding value documented. They must be maintained in good health and their reproductive activity must be properly managed. The types of crosses to be carried out need to be carefully considered based on the type of selection to be employed (mass, family or within family) and the genetic integrity of all families of offspring needs to be strictly maintained within the hatchery and under nursery culture.

There is always the possibility that artificial selection produces little or no gain. This is often the case and can be the result of there being limited genetic variation within the broodstock or the population being selected. Even if h^2 is high, in the absence of variation to exploit there can be little progress from generation to generation. One explanation for the lack of variation could be that the population was the result of an initial founder event and that many subsequent generations of significant inbreeding have resulted in high homozygosity across the genome. One obvious solution is to introduce new genetic variation into a broodstock by introducing individuals from alternative populations or broodstocks. However, care needs to be exercised to ensure that this does not risk the unintended introduction of non-native individuals into the wild.

Inbreeding, cross-breeding and hybridisation

We discussed the genetic consequences of inbreeding in an earlier chapter; now we shall consider the use of inbreeding as a tool for the production of inbred strains as part of a breeding programme.

So far in this chapter we have considered mainly the additive genetic component (V_a) of the variance of a trait. Although the dominance component (V_d) of the variance of a trait can occasionally be estimated in heritability analysis, it is not an easily predictable quantity. To remind ourselves, dominance variance is the variability of a trait due to the different combinations of alleles (as genotypes) at a locus. Dominance effects represent interactions between pairs of alleles at the same locus but genotypes are disassembled at meiosis and reassembled randomly during syngamy. Because of this, when breeding from a population we cannot guarantee the exclusive production of particular genotypes except in certain circumstances. For instance, breeding from a broodstock of animals that are all homozygous for a single fixed allele at a locus will produce all-homozygous offspring. But if some of the broodstock are heterozygotes then it is not possible to produce all-heterozygous or all-homozygous offspring.

In Chapter 4 we encountered the phenomena of inbreeding depression (Box 4.1) and hybrid vigour or heterosis (Box 4.3), and the genetic architecture associated with them: inbreeding depression–high homozygosity across the genome, hybrid vigour–high heterozygosity across the genome. These qualities are therefore directly related to dominance genetic effects. The expression of heterosis, where highly heterozygous offspring exhibit increased fitness, is obviously an important goal of hatchery production. One of the simplest ways to do this is to develop different strains or lines of highly homozygous individuals for crossing. But, of course, such highly homozygous lines risk inbreeding depression and may be difficult to grow and maintain in the hatchery. An alternative method is to cross between different natural strains, or even between closely related species. All offspring from such crosses or hybridisations will have identical heterozygous genotypes at loci that are homozygous for different fixed alleles in the parents.

Evidence for the superior fitness of cross-bred or hybrid offspring is common in agricultural crops and animals. In many species and for many traits they significantly exceed the best-performing parental line. Nevertheless, there are also plenty of examples where the crossed phenotypes are only intermediate between, or even inferior to, the two parental phenotypes. Hybridisation and crossing between strains are therefore not guaranteed to be successful in improving the performance of offspring. Nevertheless, where heterosis has been produced in cross-bred offspring some of this heterosis will remain in subsequent generations as long as random mating occurs and the effective population size is sufficiently large to avoid inbreeding.

Where a trait has a very low narrow-sense heritability (low V_a), cross-breeding can be used as an alternative to selection for that trait. However, because dominance effects are not predictable, the potential performance of an individual strain used for cross-breeding can only be assessed on the results of a number of line-by-line crosses.

In this way the mean performance of the cross-bred offspring of a particular line in relation to the average performance of cross-bred offspring from all possible lines can be assessed. This is the general combining ability of the line, while its specific combining ability is a measure of the performance of its offspring when crossed with an individual line. Currently, developments in the domestication of most aquaculture species have not reached the point at which sufficient numbers of inbred lines or strains are available to consider their combining abilities.

Hybridisation between closely related species can be seen as simply an extension of cross-breeding between strains or lines within a species. Although, strictly speaking, appropriate only for interspecies crosses, the terms 'hybridisation' and 'hybrids' are nevertheless often used rather loosely about cross-breeding within species. Part of the difficulty comes from the problem of precisely defining the blurred species boundaries in cases where extensive hybridisation can occur between 'species' in the wild. It is not the purpose of this book to indulge in semantics, but simply to point out that there are several biological definitions of the term 'species' and there are still disputes about the specific status of some common aquaculture organisms, for example, the closely related group of mussels *Mytilus edulis*, *M. trossulus* and *M. galloprovincialis*, which hybridise extensively in areas of overlap.

Hybridisation trials have been carried out in a number of groups of aquaculture organisms with varying success. When the species differ in their mechanisms of sex determination, the hybrids will often be of only one sex. The production of monosex hybrids can be a very valuable feature because it enables better control of grow-out as animals reach maturity. Where both sexes are present during grow-out, often a considerable amount of the energy provided in the feed goes towards the costs of fighting, courting and copulation. In some cases hybrids are effectively sterile and this may be advantageous.

It is also important to consider species' differences in maternal effects such as mitochondrial DNA, egg quality and brooding. For example, in hybridisations between the striped bass (*Morone saxatilis*) and the white bass (*Morone chrysops*) best results were obtained by using females of the larger striped bass crossed with male white bass. This is because the striped bass produce more eggs and the hybrid larvae from this direction of cross are larger and more hardy. However, sometimes in the real hatchery factors other than genetics control what can be achieved – adult female striped bass are not all that easy to collect nor to maintain in good spawning condition so the industry has had to compromise and use the less productive direction of cross, white bass females with striped bass males.

In oysters of the genus *Crassostrea*, hybrid crosses will usually produce larvae but in most cases these do not survive. Only the cross between *C. rivularis* and the Pacific oyster, *C. gigas*, seems to produce viable offspring, and even then there is a difference between the directions of hybridisation. Male *C. gigas* crossed with female *C. rivularis* produces feebler offspring than *C. gigas* eggs fertilised with *C. rivularis* sperm. In no cases so far have oyster hybrid offspring grown faster or survived better than either of the pure species lines in properly controlled experiments, so this does not look like a particularly rewarding avenue of further research effort. In spite of this, there is still the potential that hybridisation could bring together other qualities from the parental species, for example salinity tolerance or disease resistance, which could be more highly valued than simply growth performance.

Within the salmonids as a group, many attempts have been made to improve performance by hybridisation or cross-breeding between strains, but with mixed results. Crossing between two strains of *Oncorhynchus mykiss*, the freshwater (and effectively domesticated) rainbow trout and the anadromous steelhead trout, produced a cross-bred strain which had reduced seasonal variation in seawater adaptability compared with the pure steelhead trout. Similarly, stocking success of *Salmo trutta*, the brown trout, has been improved by cross-breeding between a domestic and a wild strain, while crosses between certain domestic strains can produce hybrids with improved growth rates. However, hybridisation trials between arctic charr (*Salvelinus alpinus*) and brook charr (*Salvelinus fontinalis*) demonstrated that hybrids were viable, but that there was no evidence for heterosis in growth or any other commercial trait. Similarly, no significant advantage in resistance to a hemoflagellate disease organism was obtained in hybrids between coho (*Oncorhynchus kisutch*) and sockeye salmon (*Oncorhynchus nerka*).

In spite of the mixed results seen in salmonid interstrain crosses and hybridisations, significant and therefore commercially valuable heterosis has been demonstrated in a number of other fish species, such as sturgeons (*Acipenser ruthenus* × *Acipenser baeri*), tilapias (Nile × red; Stirling Nile × Korean Nile; Stirling Nile × Japanese Nile) and African catfish (*Clarias anguillaris* × *Clarius gariepinus*).

These mixed results of hybridisation trials illustrate that although there is the potential for genetic improvement through hybridisation, success is very hard to predict and considerable research needs to be done on a case by case basis before commercial application is considered. Nevertheless, one prediction that can be made is that successful hybridisation between species that have different chromosome numbers, or strongly differing karyotypes (numbers of acrocentric, metacentric, etc., chromosomes), is very unlikely. Therefore a detailed knowledge of the karyotypes of potentially hybridising species is of importance for this type of development in aquaculture.

While originally only valued as ornamental fish, tilapias, or 'aquatic chickens' as they are sometimes known, are now an extremely important international aquaculture product, especially in Africa (their region of origin) and in the Far East. Using many of the methods described in this chapter, extensive efforts have been made to develop better breeds of tilapia, particularly by the International Centre for Living Aquatic Resource Management (ICLARM) based in the Philippines, where a selective breeding programme to develop genetically improved farmed tilapias (GIFT) was begun in the 1980s. Extending this approach, ICLARM has set up an International Network on Genetics in Aquaculture (INGA), which fosters research into the genetics of species such as the catla (*Catla catla*), rohu (*Labeo rohita*) and silver barbs (*Barbodes gonionotus*) in Bangladesh and India, the common carp (*Cyprinus carpio*), silver carp (*Hypopthalmichthys molitrix*) and tilapias in China, the mrigal (*Cirrhinus mrigala*) in Vietnam, the freshwater prawn (*Macrobrachium rosenbergii*) in Malaysia, and tilapias in many African and Asian countries.

Further reading

Benzie, J.A.H. (ed.) (2002) *Genetics in Aquaculture VII*. Elsevier, Amsterdam.

Doyle, R.W., Herbinger, C.M., Ball, M. & Gall, G.A.E. (eds) (1996) *Genetics in Aquaculture V*. Elsevier, Amsterdam.

Falconer, D.S. (1989) *Introduction to Quantitative Genetics*, 3rd edn. Longman, New York.

Fingerman, M. & Nagabhushanam, R. (eds.) (2000) *Recent Advances in Marine Biotechnology, Volume 4, Aquaculture, Part A, Seaweeds and Invertebrates*. Science Publishers Inc., Enfield, NH.

Gjedrem, T. (2000) Genetic improvement of cold-water fish species. *Aquaculture Research*, **31**, 25–33.

ICLARM web page: http://www.cgiar.org/iclarm/

Kearsey, M.J. & Pooni, H.S. (1996) *The Genetical Analysis of Quantitative Traits*. Chapman & Hall, London.

Lutz, G.C. (2002) *Practical Genetics for Aquaculture*. Blackwell Science, Oxford.

McAndrew, B. & Penman, D. (eds) (1999) *Genetics in Aquaculture VI*. Elsevier, Amsterdam.

Tave, D. (1993) *Genetics for Fish Hatchery Managers*, 2nd edn. Van Nostrand Reinhold, New York.

Chapter 6
Triploids and Beyond: Why Manipulate Ploidy?

An organism's ploidy is the number of copies of each chromosome set that it has. Usually gametes have a single set – they are haploid – and two gametes fuse to form a diploid zygote. However, it is viable for organisms to possess three (triploid), four (tetraploid) or more copies of each chromosome, in which case they are known as polyploids. Polyploidy can happen in nature – one example being hexaploidy in wheat – but it can also be induced. Because the eggs of most fish and shellfish are released into water before fertilisation it is relatively easy to access the maturation divisions of the egg and the early divisions of the embryo. Such access allows manipulation to create polyploid embryos and can also be used to produce diploid embryos that contain only maternal chromosomes – gynogens, or only paternal chromosomes – androgens. This is one field of biotechnological development in aquaculture that does not have its origins in agriculture, given the difficulty in accessing the egg in agricultural animals and plants.

The reader's first question might be: What is the point of ploidy manipulation? The short answer is that by manipulating ploidy we can produce sterile, unisex or highly homozygous cohorts of animals. The aquacultural value of such lines will be discussed later in the chapter; let us first consider the basic principles of ploidy manipulation. In Chapter 1 we gave an outline of the process of meiosis and it will be recalled that there is an initial reduction division (meiosis I) in which the chromosome number is halved, followed by a mitotic-type division (meiosis II) during which these haploid chromosome sets are copied into two daughter cells each. Thus, in males, four haploid spermatozoa are produced from each diploid germ cell. However, in females, both meiosis I and meiosis II produce daughter cells of such very uneven size that the smaller cells (the polar bodies) often do not complete meiosis, therefore only one haploid egg and two (rarely three) polar bodies are formed from each diploid germ cell. It is not just that fish and shellfish eggs are released into water that makes ploidy manipulation possible, but rather that the meiotic divisions have not been completed when the eggs are spawned. In the case of most molluscs, eggs are released at metaphase of meiosis I. The first polar body has yet to be produced, and an even more convenient feature is that activation by spermatozoa is required before meiosis will proceed. In most fish the eggs are released after meiosis I with the first polar body present, but further development through meiosis II is again dependent on activation by spermatozoa.

We should note that, although there is external fertilisation in most fish and molluscs, this is not generally the case in crustacean shellfish nor in gastropods such as the whelk, and that this limits the potential of ploidy manipulation in these groups. Also, among the bivalve molluscs, the flat oysters (e.g. the European native oyster, *Ostrea edulis*) retain their eggs and brood their embryos in the mantle cavity.

The basic idea behind the method of ploidy manipulation is to allow chromosome replication, but prevent cell division. In that way daughter cells have double the number of chromosomes. If we are able to suppress the cytoplasmic division of meiosis I then the resultant single cell will contain all the pairs of chromosomes and no first polar body is formed. What would normally be two haploid cells will be a single diploid cell. If, on the other hand, meiosis I is allowed to proceed as normal and cytoplasmic division in meiosis II is suppressed instead, then the chromosomes normally expelled into the second polar body are retained and a diploid cell again results. Thus, manipulating ploidy at meiosis I *or* meiosis II makes the egg diploid rather than haploid. Subsequent syngamy with the haploid male pronucleus results in a triploid embryo. In fish, meiosis II is always the division targeted for the production of triploids because meiosis I has been completed before spawning. In molluscs, while it may not initially appear to matter whether the triploids produced come from suppression of meiosis I (MI triploids) or meiosis II (MII triploids), there is a potential effect on heterozygosity, which we will consider later.

Another trick is to allow both meiotic divisions to take place, and to allow the male and female pronuclei to unite in syngamy, but to suppress the first cleavage division of the zygote. This again produces a daughter cell with double the number of chromosomes – a tetraploid. Although tetraploid fish are produced using this approach, it is far less successful in molluscs; alternative methods have been developed for the tetraploidisation of oysters and these will be discussed later. The general details of triploid and tetraploid production are illustrated in Figure 6.1.

How is it done?

So how is the cytoplasmic division of the cell suppressed? Surprisingly simply, it turns out. All that is needed is some kind of physical or chemical shock delivered at the start of division and maintained for a short period. Physical and chemical shocks are very seldom 100% effective and the proportion of eggs that respond depends upon a number of factors – principally shock magnitude and timing. Of the physical shocks, heat and cold are the easiest to administer and consist of a sudden raising or lowering of the temperature by 5–10°C from ambient. Although hot and cold shocks can produce good results in warm- and cold-water fish, hot shocks seem to produce better results in cold-water fish and vice-versa. Another physical shock method commonly used is to place eggs into a pressure chamber and subject them to pressure of up to 9000 lb in^{-2} ($\approx$ 60 megapascals or 600 bar; normal SCUBA tanks are pressurised to around 3000 lb in^{-2}). Apart from the hazards of working with such a high-pressure vessel, a further problem with this method is that the number of eggs that can be treated at any one time is limited by the volume of the pressure chamber.

Physical shocks seem to work well in fish species, but less well in molluscs. The standard method for cell division suppression in molluscs (principally developed for oysters) is chemical, involving the use of cytochalasin B (CB). CB is a fungal extract which apparently inhibits microfilament formation and development of the cleavage furrow, and therefore prevents cells from dividing successfully. As it is not very water-

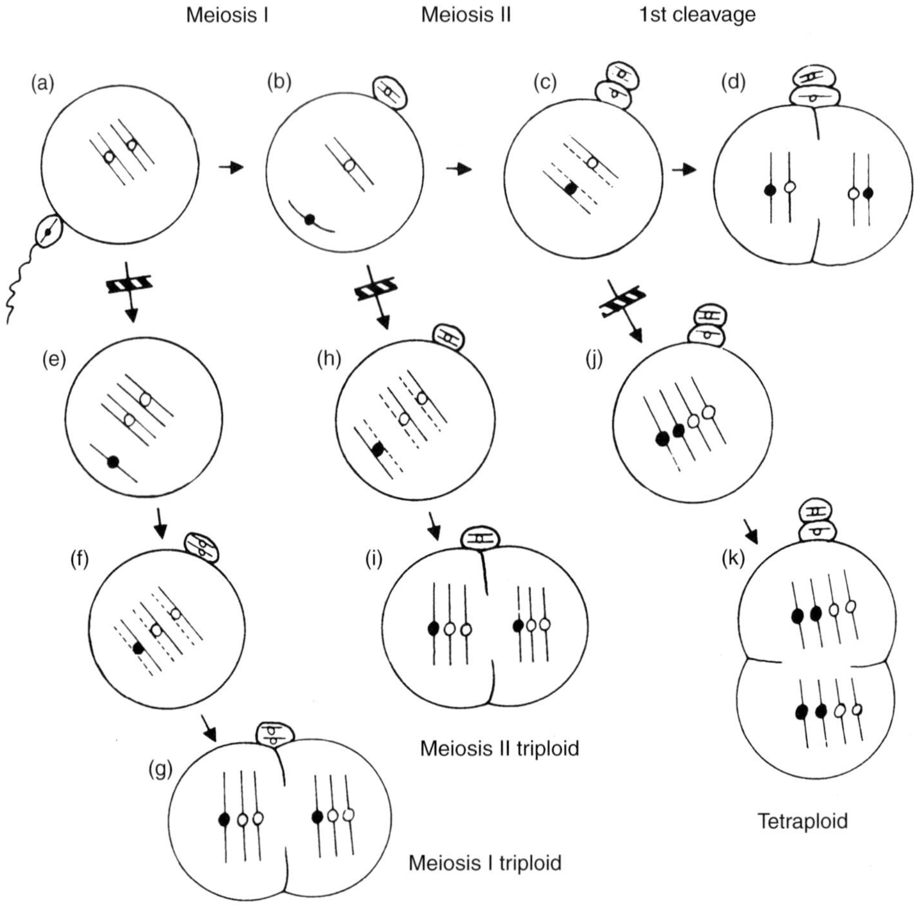

Fig. 6.1 How chromosomal manipulation works – the production of meiosis I triploids, meiosis II triploids and tetraploids. For simplicity only a single homologous pair of chromosomes is shown. (a–d) Normal development: (a) primary oocyte at release which is activated by spermatozoa; (b) meiosis I, 1st polar body extruded; (c) meiosis II, 2nd polar body extruded and syngamy occurs; (d) 1st cleavage. (e–g) Following shock at meiosis I female chromosomes are retained in the egg and a meiosis I triploid is produced. (h, i) Shock at meiosis II produces meiosis II triploids. (j, k) Shock at 1st cleavage induces tetraploids. Note that fish release oocytes which have already undergone meiosis I, so MI triploid fish cannot be produced. (From Beaumont, A.R. & Fairbrother, J.E. (1991) Ploidy manipulation in molluscan shellfish: a review. *Journal of Shellfish Research*, **10**, 1–18.)

soluble, CB is first dissolved in dimethyl sulphoxide (DMSO) which enables it to be subsequently dissolved in water. In addition, DMSO assists the chemical to enter the eggs through the cell wall (DMSO is so highly penetrative that it is also used as a cryoprotectant and as an ingredient in the spray-on pain killers used by athletes). As might be expected by its mode of action, CB is highly toxic to humans and although the amounts used are very small (up to 2 mg l^{-1}) and the application of CB takes place before the embryo begins to develop, concerns have been expressed about the dan-

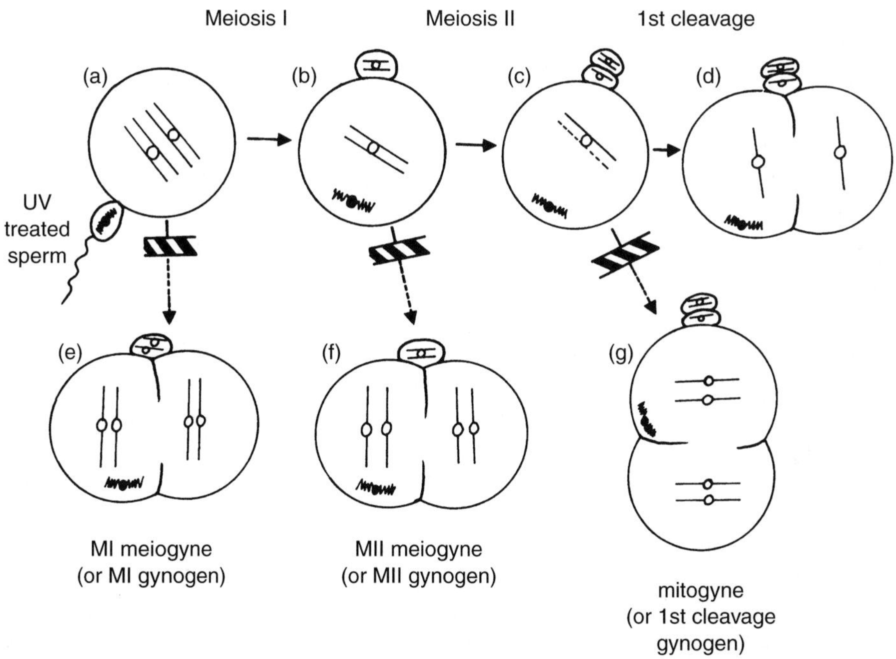

Fig. 6.2 Gynogen production. For simplicity only a single homologous pair of chromosomes is shown. The primary oocyte (a) is activated by UV-irradiated spermatozoa (sterizoa) and without any shock treatment will develop through meiosis I (b), meiosis II (c) and 1st cleavage (d) to produce a haploid embryo. With shock administered at meiosis I or meiosis II then MI (e) or MII (f) meiogynes are produced. Shock administered at 1st cleavage produces mitogynes (g). Note that fish release oocytes which have already undergone meiosis I, so MI meiogynes of fish cannot be produced. (From Beaumont, A.R. & Fairbrother, J.E. (1991) Ploidy manipulation in molluscan shellfish: a review. *Journal of Shellfish Research*, **10**, 1–18.)

ger to humans when consuming ploidy-manipulated oysters. Partly as a result of these concerns, the less toxic chemical, 6-dimethylaminopurine (6-DMAP), which is water soluble, has also been developed for use in shellfish ploidy manipulation.

Production of gynogens and androgens

The method to produce gynogens is illustrated in Figure 6.2 and involves the destruction of a complete male chromosome set together with the division suppression methods used in making triploids and tetraploids. In gynogen production, sperm are treated for a short time with X rays, gamma rays, or, most commonly, by UV irradiation. This effectively breaks up the chromosomes but does not destroy the motility or the ability of such sterile spermatozoa, or sterizoa, to activate eggs. Eggs activated by sterizoa would normally result in haploid embryos and these are invariably non-viable. However, in molluscs, any one of the three divisions – meiosis I, meiosis II or first cleavage – can be suppressed in order to double the chromosome number, while in fish, meiosis II and first cleavage can be targeted to ensure diploidy. In all cases, the diploid embryos produced contain only the chromosomes of the female parent. Because there is no contribution from male chromosomes inbreeding is high, but the level of inbreeding in gynogens will depend on which division is targeted. Doubling the chromosome number at first cleavage results in homozygosity at every locus in the genome. That is not to say that all the embryos are identical clones, but each individual is homozygous at all of its loci and thus 100% inbred. These mitotically-produced individuals are called mitogynes and, because all loci are homozygous, any deleterious recessive alleles are exposed with consequent reduction in viability. Gynogens produced by suppression of meiosis I or meiosis II are called meiogynes and they will not be entirely inbred because of recombination events occurring during meiosis. We will consider this in more detail later.

Androgens are individuals whose chromosomes are entirely paternal. They can be produced by two methods, both of which require that eggs are irradiated to destroy their chromosomes. The first uses normal haploid sperm to fertilise the irradiated eggs and then the first cleavage division is suppressed to produce a diploid embryo. The second employs diploid sperm – the gonad product of a tetraploid male – to fertilise the irradiated egg and diploid embryos are produced without further treatment.

Identification of ploidy change

Production of triploids by physical or chemical suppression of divisions seldom results in 100% triploids and production of tetraploids is even less certain. For this reason, hatcheries need to establish how successful their triploid or tetraploid induction efforts have been. A balance must be struck between the proportion of triploids in a batch and the costs of ensuring triploidy induction. There are a number of ways in which the ploidy of individual fish or shellfish can be identified and the quickest method is to use a flow-cytometer, which measures the amount of DNA-specific stain taken up by nuclei of individual cells in a sample. A graphical display allows calculation of the proportion of individuals in the sample that are triploid and the proportion that are diploid (or of other ploidies) by measuring the areas under the peaks coinciding with the amount of DNA present in the nuclei of each ploidy type. The main problem with this method is that the price of a flow-cytometer is prohibitive for most hatcheries. Nowadays, however, there are companies that have specialised in the production of triploids of one or two of the main commercial aquaculture species and they will have their own flow-cytometers. Smaller producers can buy in ready-produced triploid seed or fry. In addition, research institutions and private companies with these facilities are not averse to providing a costed service for smaller-scale operators who may wish to carry out their own ploidy modification work.

Other methods of identifying triploids and tetraploids include direct chromosome counting, sizing of nuclei of cells from blood or other tissues and scoring genetic markers. Chromosome counting requires experience and can be quite time consuming, but has the advantage that it is a direct method. Using blood for nuclear sizing is cheap, rapid and effective and can be non-destructive but nuclear sizing of the cells of other tissues requires the sacrifice of the sampled animals. Use of genetic markers such as allozymes for ploidy confirmation relies on the fact that some polyploids will exhibit more than two alleles at highly polymorphic loci. For example, a triploid could express three alleles, with the genotype ABC. In theory, a triploid heterozygote with only two alleles (e.g. AAB) could be distinguished from a diploid (AB) by the stronger staining of the band produced by the double allele, but in practice this is not very reliable. However, the use of microsatellite loci is a great improvement on allozymes because (being highly heterozygous) they are more likely to show three bands and (because minute amounts of DNA are required) they can be scored by non-destructive sampling.

Confirming that diploid gynogens or androgens have been successfully produced is a bit more tricky than confirming the presence of triploids or tetraploids. Obviously, there is no difference in chromosome number, nuclear size or DNA content between normal diploids and gynogens or androgens. Here we will consider gynogens, but similar methods apply to androgens. Remember that gynogens are produced by activating eggs with sterizoa and then using a shock to suppress a cell division (meiosis I, meiosis II or first cleavage). One method that provides indirect confirmation of gynogen production is to use two controls without the shock to suppress cell division – one using untreated spermatozoa and the other using sterizoa. If everything has worked properly (sperm irradiation and the shock method) then the untreated spermatozoa control will produce normal diploid offspring while the sterizoa control will produce haploid embryos that do not survive. If these controls show the expected results then any surviving embryos from the fully treated group should be gynogens.

Sometimes the spermatozoa of a closely related species can be used to fertilise eggs, producing non-viable hybrids. In this situation sterizoa of the closely related species can be used in gynogen production. Given that hybrids are non-viable, any surviving offspring must be gynogens.

Alternatively, genetic markers can be used to confirm individuals as gynogens. For example, albinism is a recessive trait and spermatozoa from a male trout which is homozygous for pigment production (AA) can be used with eggs from an albino (aa) trout. Gynogens will all be albino (aa) and normal diploids will be pigmented (Aa). Allozyme or microsatellite loci can also be used to identify gynogens when the allele(s) present in the spermatozoa are different to the allele(s) present in the egg.

Triploids

Production of triploids is the most common target of ploidy manipulation attempts and they have a variety of uses in aquaculture. Triploids have been produced, at least experimentally, in almost all commercially aquacultured fish species, including carps, catfish, *Tilapia* and salmonids, and in most molluscan shellfish groups such as oysters, clams, scallops and abalones. One important reason for producing triploids is that they are sterile. During early gametogenesis, germ cells begin the process of meiosis and it could be that attempts to pair up the chromosomes in early meiosis are impeded by the fact that there are three homologous chromosomes rather than two.

Whatever the reason, gonad development is always much reduced in triploids. Some eggs and spermatozoa *can* develop in triploids, but these are seldom of normal ploidy and are usually aneuploid (missing chromosomes or parts of chromosomes). Triploid sterility means that, as triploid fish or shellfish reach maturity, energy that, in diploids, would go to developing gametes is available for somatic growth. As they get older, therefore, triploids effectively grow faster than diploids. Whether this potential for increased somatic growth is realised as a reduced time to market size depends on a number of factors, not least the species being cultured. Triploid fish and shellfish generally demonstrate a clear growth advantage over diploids during annual gametogenesis, but there may be more than one year of maturity before market size is reached and the advantage from the first year may not be carried over to the next. Similarly, superior growth performance of triploids in many fish species has only been demonstrated for a small part of their life histories.

Apart from the actual or potential increase in somatic growth, there are other advantages that accrue from the sterility of triploids. Sterility makes feasible the aquaculture of non-native species or genetically depauperate hatchery-produced stock, either of which might otherwise cause adverse environmental impact if they or their gametes were to escape into the wild. Before beginning such an enterprise it is necessary to confirm, first, that all individuals are triploid and, second, that even if they produce some gametes these will never produce offspring. Considering the first requirement, methods are needed to ensure 100% triploid production and also to assess the ploidy of every individual that is to be introduced. Rather than try to perfect other triploidy induction methods, the development of tetraploids as broodstock has been the favoured option. A tetraploid individual will produce diploid gametes and these, when combined with haploid gametes from a normal diploid, produce 100% triploids. Triploids produced by this method have been called interploid triploids. Generally tetraploid males are more valuable because the diploid spermatozoa produced can be used to fertilise the eggs from a large number of females. Even though this method is quite certain to produce only triploids, regulatory bodies are wise to insist on certification that all individuals for importation are triploid, based on flow-cytometry or some other proven method, and that any gametes produced from triploid individuals are effectively non-viable.

The licensed importation of non-native, triploid grass carp into many states in the USA was only allowed when research had established that, even after induction of gametogenesis by hormonal injection, almost all the spermatozoa released by these triploids were abnormal. Calculations indicated that the probability of triploid grass carp having fertile offspring was so low that they could be considered to be sterile for all practical fishery management purposes.

Another example of the importance of regulation and monitoring of the introduction of triploids of non-native species concerns triploid Pacific oyster (*Crassostrea gigas*) on the east coast of the USA. The backgound to this is that the American oyster *Crassostrea virginica* has suffered severe declines over recent decades, mainly caused by two protozoan parasites – dermo (*Perkinsus marinus*) and MSX (*Haplosporidium nelsoni*). Several eastern USA states had considered the possibility of introducing the Pacific oyster, which is not affected by these diseases, as an alternative for the local oyster culture industry. A trial introduction programme of triploid *C. gigas* was instigated, in which the triploid nature of every individual was confirmed by flow-cytometry of blood samples. After a year or so, as oysters were retested for ploidy confirmation, it was discovered that some oysters (or at least some tissues from some oysters) had reverted to diploidy. Obviously this was a serious blow to the plans to use triploid oysters in this way, and further research is continuing to explore this unexpected phenomenon.

In spite of uncertainties concerning reversion to diploidy, there is one area where triplod *C. gigas* have been a commercial success. Pacific oysters (*C. gigas*) were first introduced from Japan into the Pacific north-west of the USA in the middle of the last century and soon became naturalised. They have formed the basis of a significant hatchery-based aquaculture industry supplying Pacific oysters to the market at a time when production of the native American oyster (*C. virginica*) was in decline. However, there is a problem with Pacific oysters which is less significant in *C. virginica*: they become unmarketable when they reach sexually maturity in the summer. The reproductive tissues ramify throughout the oyster's body and its glycogen stores are converted into gametes. This changes the normal sweet flavour of the oysters and significantly reduces the quality of both taste and texture. In triploid oysters little gametogenesis takes place and stored glycogen remains unconverted and they therefore provide a year-round marketable product. With the successful recent development of tetraploid Pacific oysters, commercial supplies of 100% triploid oyster seed are now readily available and most commercial production of Pacific oysters in the USA is now of triploids.

A further potential value of triploidy in bivalves concerns the production of artificial pearls from pearl oysters (*Pinctada* spp.). The method requires that a 'seed' pearl is implanted into tissue close to a vestigial loop of the gut, but it is a tricky operation and usually results in some mortality and some implant rejections. In ripe individuals, the gut is surrounded by gonadic material which adds to the difficulty of implantation, so sterile triploids could be used to reduce mortality and implant rejections.

In salmon aquaculture, triploids have been proposed as a method to overcome problems associated with grilsing. Grilse are fish (both male and female) that mature after only one year at sea, leading to a deterioration in flesh quality and a transfer of energy from somatic to gonadic growth as maturity approaches. Identifying, sorting and trying to market grilse creates practical difficulties, and the use of sterile triploid salmon was an approach considered by the industry. However, strains of salmon that have been artificially selected for late maturity are now available and these have become the preferred means of reducing grilsing in salmon farms.

The process of triploidisation can also be of value when interspecific hybrids are produced. For some reason, possibly the presence of a double set of chromosomes from one or other parent species, interspecific hybrid triploids can often outperform their diploid counterparts. Indeed, in some circumstances triploidisation of hybrids may be the only way to produce survivors of interspecific crosses.

For some shellfish, such as sea urchins and some scallops (in Europe), the roe or gonad is the important market product. In these cases there is obviously no commercial advantage to producing triploids.

	Without recombination	With recombination
Diploids	AC or BC	AC or BC
Meiosis I triploids	ABC	AAC, BBC or ABC
Meiosis II triploids	AAC or BBC	ABC

Fig. 6.3 Genetic consequences of triploidy induction in relation to heterozygosity. Potential genotypes of diploid and triploid offspring from a female heterozygous for alleles A and B, mated with a male homozygous for allele C. Note that fish release oocytes which have already undergone meiosis I, so MI triploid fish cannot be produced.

Although the advantages that accrue to triploids through their sterility are the most important ones from the aquaculture point of view, we should also consider the question of the potential effect of increased overall heterozygosity of the genome of triploids compared with diploids (Fig. 6.3). Genetic theory is steeped with the notion that heterozygosity is an advantage over homozygosity, and that this is linked with the phenotypic phenomena of hybrid vigour and inbreeding depression. Accepting this, there is nevertheless a difficulty in mapping these effects back to heterozygote advantage at more than a very small proportion of single gene loci in the genome as is allowed under neutral theory. In-depth discussion of this topic is outside the scope of this volume, but we must consider that the increased heterozygosity across the genome in triploids could be similar to the increased heterozygosity of hybrid genomes. In that case, we should expect to find evidence of increased performance in many quantitative traits, and not just in growth, in triploids. Leaving aside the advantages due to sterility that we have already considered, performance in triploids (usually measured as increased growth at larval or juvenile stages) is sometimes, but by no means always, better than that of diploids. It is very seldom worse.

To confuse matters further, there is the question of how the cells of triploids cope with having a larger nucleus than the cells of diploids. In plants, where cell walls are quite rigid, increased ploidy does result in increased cell size, but is this the case in animal cells? And if triploid cells are larger, are there as many of them in a triploid as in a comparable-sized diploid individual? If there are the same numbers of cells in diploids and triploids, but triploid cells are larger, then this could account for any apparent faster growth. This is called polyploid gigantism. On the other hand, the relative proportions of nucleus and cytoplasm in cells will have arisen by the process of natural selection, and therefore any deviation from these proportions might put an individual at a selective disadvantage. In that event we might expect triploid fish to underperform in comparison with diploids. While there is no evidence of major differences in the biochemistry or physiology of triploids compared with diploids, there is still much to learn about the more subtle consequences of triploidisation in fish and shellfish.

Tetraploids

The principal value of tetraploids is that they can be used as a source of diploid gametes in the production of interploid triploids and androgens. Tetraploid fish have been successfully produced by targeting the first cleavage division, but this approach has not been productive in molluscs. It is possible that the differences in early embryonic divisions in these two groups may account for this. In fish the early cleavage divisions are equal. In contrast, the two-cell stage in bivalve molluscs consists of two unequally sized nucleated cells with a cytoplasmic 'polar lobe' extruded from the larger of them. The polar lobe is resorbed back into this cell during second cleavage. Suppressing first cleavage, if it interferes with polar lobe formation, is likely to have considerable consequences on early embryonic development and might explain the non-viability of tetraploids produced in this way.

Because of the failure of suppression of first cleavage to produce viable tetraploid molluscs, researchers have sought an alternative approach. Small numbers of tetraploid broodstock of the Pacific oyster have now been produced by suppressing the meiosis II division in (rare, often aneuploid) eggs from triploid oysters and fertilising them with ordinary haploid spermatozoa. A number of different ploidies, including a few tetraploids, are produced by this method and, unlike tetraploid embryos induced by suppression of first cleavage, some of these tetraploids survive. It is suggested that the extra cytoplasm present in a triploid egg is needed to support a tetraploid nucleus and that this explains why these tetraploids are viable.

Adult male tetraploid fish and oysters have now been used successfully to produce interploid triploids, although diploid spermatozoa are less active and less viable than normal haploid spermatozoa. Eggs from female tetraploid fish have also been successfully crossed with normal haploid spermatozoa.

Obviously, using diploid sperm from tetraploid males to fertilise diploid eggs from tetraploid females will result in all-tetraploid offspring, and tetraploid broodstock lines have been developed for rainbow trout (*Oncorhynchus mykiss*) and some other fish. It is possible that in tetraploids some of the potential dangers of the homozygous exposure of deleterious recessive genes as a result of inbreeding in an enclosed broodstock might be ameliorated by the fact that there are four copies of each chromosome. On the other hand, nothing is known about the long-term genetic, physiological or biochemical consequences of continual inbreeding of a tetraploid stock.

Gynogens and androgens

Diploid gynogens are individuals that contain two copies of the maternally inherited chromosome set. One of the most useful features of gynogens is that in species where there is an XY sex-determining system, all gynogens will be XX and therefore all female. Monosex populations of fish can also be produced by hormonally-induced sex-reversal, but gynogen production avoids the use of chemicals and produces genotypic, rather than 'phenotypic', females. If some of the all-female gynogens are then hormonally sex-reversed to phenotypic males, they can then be mated with other gynogens to produce all-female offspring without having to go through the process of gynogenesis again.

Simple chromosomal sex determination in fish may be by the XY system, where the female is the homogametic sex (XX), or by the WZ system, where the male is the homogametic sex (WW). However, sex determination is usually more complex than a simple chromosomal system as several autosomal gene loci are also involved. Gynogens can be used to explore whether a simple chromosomal sex-determining mechanism is present in a species: if half are male and half are female this would point to a WZ system; if all are female the XY system is indicated. The sex-determination mechanism of molluscs is not well understood, but certainly does not consist of a simple chromosomal system. Indeed, many bivalve molluscs are hermaphrodites and some, including oysters, are sequential hermaphrodites, usually changing from male to female. So the study of gynogens is not expected to be much help in elucidating the sex determination mechanisms in bivalve molluscs.

A further important value of gynogens is that they are strongly inbred and have the potential to provide the basis for inbred lines in just a single generation. Mitogynes are expected to be 100% homozygous at all their gene loci while meiogynes will be less inbred, and the level of homozygosity at particular loci will depend on the frequency with which those loci are involved in recombination events (Fig. 6.4). Such high levels of inbreeding result, as expected, in inbreeding depression. Gynogens generally have higher mortality and slower growth than normal diploids, but if they survive to maturity they can provide an inbred line for later outcrossing. As discussed in Chapter 5, inbred lines that are homozygous at many or all of their loci can be crossed to produce highly heterozygous F1 offspring; the value of F1 hybrids is well proven and their use is extremely common in agriculture.

There is a further potential use of gynogens. If it is possible to produce mitogynes that reach maturity and produce eggs, and these eggs can again be treated to produce mitogynes, then these offspring will be genetically identical clones of one another. They will be homozygous for the same allele at every locus. The experimental value of such clones of a sexually reproducing species is high.

In a further twist, it has proven possible to develop 'supermale' tilapia individuals that have two Y chromosomes but no X chromosome (Fig. 6.5). These YY supermales produce only male offspring and therefore provide the complement to female monosex production.

	Without recombination	With recombination
Meiosis I	AB	AA or BB
Meiosis II	AA or BB	AB
First cleavage	AA or BB	AA or BB

Fig. 6.4 The genetic consequences of gynogenesis induction on heterozygosity. Eggs are from a female heterozygous for alleles A and B. Note that fish release oocytes which have already undergone meiosis I, so MI meiogynes of fish cannot be produced.

Fry

↓

Hormonal feminisation

↓

XY *phenotypic females*

↓

Eggs

↓

Activation with irradiated sperm

↓

Heat shock to suppress meiosis II

↓

Gynogens : 50% XX females and 50% YY males

↓

Cull all females, remaining fish are YY supermales

Fig. 6.5 Method for the production of YY supermale tilapia.

Diploid androgens differ from gynogens in that both chromosome sets are obtained from the male parent. The uses of androgens are similar to those of gynogens. In fish with a WZ sex-determination system, androgens will all be male (WW) and this therefore enables monosex production. Alternatively, in fish with an XY sex-determination system, half of the androgenic offspring will have two Y chromosomes and, if they survive to sexual maturity, will be YY supermales.

The reader should note that although there are all these possibilities for the use of androgens and gynogens, not all have yet been realised. Research has been mainly directed at developing and refining methods for their production; their use in commercial hatcheries is not yet routine.

Further reading

Beaumont, A.R. & Fairbrother, J.E. (1991) Ploidy manipulation in molluscan shellfish: a review. *Journal of Shellfish Research*, **10**, 1–18.

Beaumont, A.R., Fairbrother, J.E. & Hoare, K. (1995) Multilocus heterozygosity and size: a test of hypotheses using triploid *Mytilus edulis. Heredity*, **75**, 256–266.

Benzie, J.A.H. (ed.) (2002) *Genetics in Aquaculture VII*. Elsevier, Amsterdam.

Doyle, R.W., Herbinger, C.M., Ball, M. & Gall, G.A.E. (eds) (1996) *Genetics in Aquaculture V*. Elsevier, Amsterdam.

Fingerman, M. & Nagabhushanam, R. (eds) (2000) *Recent Advances in Marine Biotechnology,*

Volume 4, Aquaculture, Part A, Seaweeds and Invertebrates. Science Publishers Inc., Enfield, NH.

Lutz, G.C. (2002) *Practical Genetics for Aquaculture*. Blackwell Science, Oxford.

McAndrew, B. & Penman, D. (eds) (1999) *Genetics in Aquaculture VI*. Elsevier, Amsterdam.

Tave, D. (1993) *Genetics for Fish Hatchery Managers*, 2nd edn. Van Nostrand Reinhold, New York.

Chapter 7
Genetic Engineering in Aquaculture

What do we mean by 'genetic engineering'? You may think that the cutting and splicing of DNA and alteration of chromosome numbers that we have already discussed should qualify as genetic engineering; however, the term is used to refer specifically to the transfer of genes between species. Given what was discussed in Chapter 1 about the boundless quantities of genetic variation *within* species, why should we wish to transfer genes *between* species? The short answer is time. Why go through the process of selection for a trait when we can take the gene from another species and drop the trait directly into our organism's genome? Selective breeding is still preferable where the trait is not too difficult to achieve this way, but genetic engineering is really the only commercially feasible way of bridging large gaps between an organism's natural characteristics and what the aquaculturist wants. For example, much of Canada's coastline is too cold for salmon aquaculture. Artificial selection might eventually produce a cold-tolerant strain, but only after many generations of careful breeding, which would be beyond the scope of an aquaculture business. Genetically modifying salmon to produce the antifreeze protein from winter flounder could extend their range instantly. Alternatively, it is possible that aquatic organisms could be engineered to produce novel or medicinal substances with fewer animal welfare problems than when mammals are used.

The DNA construct

The initial step in transgenesis (creation of a genetically engineered organism) is to design and build a DNA construct. Genetic engineering generally has the aim of inserting DNA which enables the genetically modified organism (GMO) to produce a protein in a specific tissue at appropriate times and at a sufficient concentration. The construct therefore must consist of the DNA to code for the protein of interest (the transgene) together with DNA coding for regulatory elements that direct gene expression (the promoter).

The transgene

The most commonly employed genes so far in fish transgenesis are growth hormone (GH) genes – the specific GH genes of humans, rats, cattle, chinook salmon, rainbow trout, tilapia and sea bream have all been used in attempts to increase growth rates and thus reduce time-to-market size. Another gene which is of special relevance to aquaculture in polar regions is the antifreeze protein gene which is found in fish such as ocean pout or winter flounder that can live in subzero temperatures. As mentioned above, if other species could be genetically modified to produce an antifreeze protein then this could extend the area available for aquaculture. Transgenesis is also being developed as a way to induce immunity to disease-causing organisms in fish. It is possible to induce immunity somatically by injecting DNA that codes for viral protein directly into muscle tissue, but further research on such genetic vaccines will be required before veterinary approval. Inserting the gene for the viral protein into the germline, however, would mean that lines of fish could be produced with built-in immunity to a particular virus.

As far as aquacultured invertebrates or plants are concerned, there does not appear to be any current commercial transgenesis. There is research interest, however, particularly on the use of genetic modification to produce disease-resistant bivalve molluscs.

Finding the gene that codes for a protein of interest can be a long and laborious process. Fortunately, more gene sequences of more species are being identified every day (e.g. the SALMAP project, Chapter 4) and these can provide the basis for the design of probes to search for similar genes in other species. Alternatively, if the amino acid sequence of the protein is known it can be used to design a DNA probe for the gene.

When a gene is developed as part of a construct for transgenesis it is not usually taken directly from the donor's DNA. This is because, as we noted in Chapter 1, gene sequences are interrupted by regions of DNA that are not part of the coding sequence. These introns are spliced out during production of mRNA. If mRNA is extracted from cells or tissues where the gene is expressed, then complementary DNA (cDNA) can be synthesised from the mRNA by the enzyme reverse transcriptase. This cDNA can then be inserted into a plasmid vector and cloned into *E. coli* in the normal way (Chapter 2, Box 2.1) to produce a cDNA library. The cDNA library is then screened using a probe based on the DNA sequence of the gene (or even the sequence from a related species or group, if the sequence is unknown for the species in question) to identify the matching cDNA. The cDNA is then snipped out of the plasmid and spliced to the promoter sequence.

In some cases it appears that certain introns are required for the full expression of the gene. This implies that some introns may have a regulatory function even though the sequence is not translated.

The promoter

There are estimated to be from 30 000 to 100 000 genes in the vertebrate genome but only about 10%, mainly housekeeping genes involved in simple cellular functions, are switched on and expressed in every cell all of the time. In addition, perhaps a few thousand more genes are involved in directing activities specific to a cell's particular function. So the majority of genes are not normally switched on in a cell and a critical requirement of transgenesis is that the gene inserted into the GMO be expressed. For this reason any gene sequence in a construct is accompanied by a promoter. The promoter consists of DNA sequences that direct transcription and translation of the gene into protein. The promoter sequence regulates when, where and how much expression of the gene will occur. Early experimental work on fish transgenesis used promoters from avian (Rous sarcoma virus, RSV) or primate viruses (simian virus, SV40; cytomegalovirus, CMV) but recently, efforts have been made to design an 'all-fish' promoter from fish genomes rather than the genomes of other groups. Examples of the genes and promoters that have been used for transgenic research into aquacultured species are given in Table 7.1.

Once the DNA construct has been formed by splicing together the gene and the promoter, it must be amplified to produce the billions or trillions of copies needed for each transgenesis attempt. The method used is to insert the construct into a plas-

Table 7.1 Major aquaculture species which have been genetically engineered with examples of the constructs used

	Construct	
Species	Gene	Promoter
Salmon	hGH, bGH, rGH	mMT-1
	csGH	opAFP
	sbGH	wfAFP
	coIGF	csMT-1
Trout	hGH	SV40
	rGH	mMT-1
	sbGH	cβ-actin
	INHV-G protein	CMV-tk
	cα-globin	cα-globin
Tilapia	hGH, rGH	mMT-1
	rGH	cβ-actin
	tiGH	RSV
	tiGH	CMV
	wfAFP	RSV
Catfish	hGH	mMT-1
	rtGH, csGH	RSV-LTR
Carp	hGH	mMT-1
	rtGH	RSV-LTR

Genes transferred: GH = growth hormone (h = human, b = bovine, r = rat, cs = chinook salmon, sb = seabream, ti = tilapia, rt = rainbow trout), coIGF = coho salmon insulin-like growth factor, IHNV-G-protein = infectious hematopoietic necrosis virus G-protein, cα-globin = carp α-globin, wfAFP = winter flounder anti-freeze protein.

Promoters used: mMT-1 = mouse metallothionein, opAFP = ocean pout antifreeze protein, wfAFP = winter flounder anti-freeze protein, csMT-1 = chinook salmon metallothionein, SV40 = simian virus 40, cβ-actin = carp β-actin, CMV-tk = cytomegalovirus with herpes simplex virus thymidine kinase, cα-globin = carp α-globin, RSV = Rous sarcoma virus, CMV = cytomegalovirus, RSV-LTR = Rous sarcoma virus with long terminal repeat.

mid which is replicated in *E. coli* by standard cloning methodology (Chapter 2, Box 2.1). In most experimental trials, and certainly in any commercial application, the plasmid DNA is removed before transgenesis because this prevents the insertion of prokaryote plasmid DNA into the target organism. The potential effects of prokaryote DNA on eukaryotic transcription and translation processes include methylation (replacement of hydrogen by a methyl group, CH_3, in a DNA base) but these effects are not predictable and must therefore be avoided. Of course, there are hundreds of millions of bacterial genomes normally in intimate contact with organisms in their gut, gills and exterior surfaces, but these intimate associations have evolved, and are continuing to evolve, by natural selection and can therefore be regarded as 'normal' and environmentally balanced.

Once the DNA construct has been created, amplified and purified out of the plasmid, it can be delivered into the eggs of the target organism.

Transgene delivery

Several transgene delivery methods have been used but none give more than limited success. For this reason, special reporter genes have been used. These are simply genes that are very obvious if they have successfully been incorporated into the host genome – for instance, they code for proteins which cause the embryo to turn blue or fluoresce green. This makes it easier to compare the success of different delivery methods and develop the best technique for delivery of the true target gene.

Microinjection

Microinjection was first developed for use on mice and it is the most common transgene delivery method used for fish, though other methods seem to be more promising for molluscs. It simply involves using remote-control levers to operate a micropipette tipped with a very fine glass needle. Individual fertilised eggs are held steady by suction against a blunt ended tube and the micropipette needle is inserted into the egg (Fig. 7.1). A small volume of buffer (usually 1–2 nl) containing a high copy number (10^6–10^7 copies) of the transgene construct is then injected into the egg, the needle is removed and the egg incubated in the normal way. Some of the difficulties inherent in the microinjection of mouse eggs, such as working *in utero* with minute eggs, are removed when working with fish. Nevertheless, although fish eggs are abundant, large and easily accessible they do have characteristics that make microinjection difficult. Fish eggs are surrounded by a chorion that rapidly hardens after fertilisation, making it difficult to penetrate. Partly owing to the hardening of the chorion, it is also extremely difficult to identify the position of the chromosomes in fertilised eggs. Being able to locate the injected DNA as close as possible to the region where the chromosomes are located is an important criterion, because otherwise the integration of the novel DNA into the chromosomes of the GMO is far less

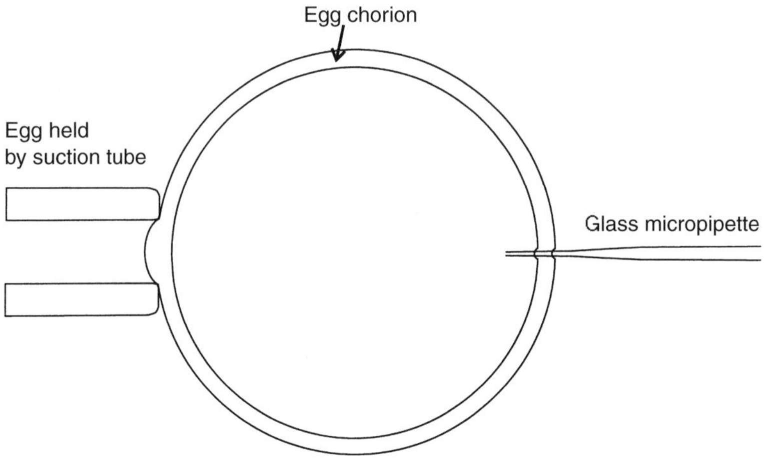

Fig. 7.1 Microinjection of DNA construct into a fish egg.

likely. However, without special *in vivo* fluorescent staining and the use of a UV light source which enables precise localisation of the male and female pronuclei, injection of construct into the pronuclear region is bound to be rather hit and miss. Attempts have been made to remove the chorion from eggs before microinjection, but this turned out to be a slow and labour-intensive procedure which did not significantly improve success. In some species, the egg has a micropyle through which microinjection is feasible.

In contrast to mice, the time from spermatozoa activation to first cleavage in fish is quite short and early embryonic cell division is rapid, which means there is only a short window of opportunity for microinjection into the fertilised egg or early embryo. Even so, depending on the species of fish, an experienced operator with logistical support may be able to inject several hundred eggs before the cleavage divisions are too advanced. Ideally the transgene should be injected at the single cell stage to ensure integration in all the cells of the GMO. However, even when this is done, integration (if it happens at all) often only takes place after some cleavage divisions have occurred. This produces mosaic embryos where the transgene is present in some cells but not others. Mosaicism is a very common outcome of transgene delivery by the microinjection method.

In general, the injection method is rather inefficient. Injection can cause damage that affects embryonic survival and can result in quite high mortalities. Often fewer than 50% of eggs which have been microinjected express the transgenic DNA in early embryos and expression can decrease significantly after a few days. Usually fewer than 1% of microinjected salmonid eggs develop through to the adult stage with the transgene integrated into the genome, though better results can be obtained experimentally with smaller fish species such as zebrafish (*Danio rerio*) and medaka (*Oryzias latipes*).

Electroporation

Electroporation is a method which was first developed for work with cells in tissue culture and it involves subjecting the cells to a short burst of electrical impulse. When cells are treated like this, their cell walls become temporarily much more porous and larger molecules than usual can pass through. Cells to be treated are suspended at high concentration in a solution of high copy number DNA construct and held in a 1–2 ml container with flat electrodes on each side. The transient current is passed and DNA construct molecules pass through the cell membrane. The electrical pulse can have a range of voltages and different rates of decay after the pulse and these are characteristics that are varied in order to produce optimum electroporation results. A major advantage of electroporation over microinjection is that there is no need to handle and manipulate eggs individually.

Electroporation has been tried on fish eggs, but the difficulty is that the eggs are quite large and have a chorion. Even after chorion removal, which is time consuming and therefore reduces the number of eggs that can be treated within the appropriate time window, the results have not encouraged further development. The stresses on embryos seem greater with this electrical method than the cell puncture used in the injection method, since embryo survival following electroporation is poor. While there are problems with fish eggs, the method has great potential for transgenesis in molluscs and successful experimental trials have been conducted with abalones and oysters. The eggs of most commercially important molluscan shellfish are rather small – less than 100 μm in diameter – which makes it quite tricky to micromanipulate individual eggs. On the other hand, there is no development of an impermeable layer following fertilisation and the egg cell wall remains unprotected by any kind of chorion. For these reasons electroporation has greater potential than microinjection for transgenesis in molluscan shellfish.

Sperm-mediated transfer

When thought about, it's obvious. What is the one thing that is guaranteed to get into the egg during fertilisation and deliver a package of DNA directly to the female pronucleus? The spermatozoon, of course. It turns out that DNA binds readily to the outer coat of spermatozoa, so at first sight this looks like a perfect method of getting novel DNA into the egg. Encouraging though this prospect seems, it is evident that there are a number of pitfalls with this method. First, in most cases it is only a single spermatozoon that enters the egg and reaches the pronucleus. Therefore the number of copies of the DNA construct entering the egg or reaching the target is quite small when compared with the numbers of copies employed in electroporation or microinjection. Secondly, it is now clear that there are probably mechanisms in place in the host to prevent the ingress of foreign DNA by this process. Spermatozoa will often come into contact with extraneous DNA (from cellular debris or bacteria) before fertilisation so it is not surprising that mechanisms have evolved to prevent such DNA from integrating into the host genome. Attempts have been made to electroporate spermatozoa with DNA construct before fertilisation; although this approach has not been extensively explored, early work suggests that the difficulties of the methodology outweigh any benefits.

Biolistics

In plants, where there is a tough cell wall, experimenters have resorted to brute force to get foreign DNA into the cells. The method is called biological ballistics, or biolistics, and involves coating microscopic particles, usually of gold, with DNA construct and explosively firing these particles directly into the cell through the cell membrane. This method has been tried with fish, sea urchins and oysters and results suggest that viable transgenic embryos can be produced. However, there is currently no evidence that biolistics offers a more effective or more efficient method of producing transgenic fish than microinjection.

Viral vectors

An alternative gene transfer method is to use as a vector a virus that has been rendered defective. The transgene is spliced into the defective virus that is nevertheless able to infect host cells and induce the replication of the transgene within those cells. However, there are obvious dangers involved in engineering any virus to be *too* good at getting into a wide range of species.

Lipofection

Synthetic lipid vesicles containing encapsulated DNA can be taken up directly by animal cells and this lipofection method has been tried with fish eggs. As with sperm-mediated transfer, initial expression of foreign DNA was followed by a rapid loss of expression suggesting that the novel DNA may have been destroyed by the host embryos.

All that these delivery methods do is to get the construct into the egg. It must then become incorporated into one or more of the chromosomes of the host. If it just remains floating around in the cytoplasm of cells, or as a fragment of DNA in the nucleus not attached to a chromosome, it will probably not be replicated and will be lost during cell division. Fortunately, a very small proportion of inserted construct becomes integrated spontaneously – which is what is currently relied upon in commercial transgenesis – but methods are being developed to increase transgene integration success.

Transgene integration

A major difficulty with the integration of a transgene is targeting its position on a chromosome. Chromosomes are not homogeneous throughout their length but consist of heterochromatic and euchromatic regions, and genes in heterochromatic regions are seldom switched on. If the transgene gets incorporated in such regions then it is unlikely to be expressed due to this position effect. Furthermore, some research has shown that transgenes that had become successfully integrated and expressed in F0 and F1 fish were not expressed in later generations. Assuming the transgenes were still integrated, their expression must have been switched off in some way.

Three main experimental approaches are being used to try to get transgenes into the nucleus and to locate them more precisely on the chromosomes. First, many viruses have proteins containing special amino acid sequences that assist the viruses to get into the nucleus where they can replicate. These are called nuclear-localising sequences (NLS) and such sequences can be added to the medium containing the transgene construct to assist with entry into the nucleus.

Secondly, a pseudotyped retrovirus (murine leukemia retrovirus with one of its genes replaced by one from the vesicular stomatitis virus) that encodes a special integrase protein can be used. Once inside the cell the integrase protein is produced and this can assist with integration of the transgene DNA into the chromosome. However, there are significant risks attached to working with these viruses because they can infect any cell, including those of the human experimenter! Far better to try to isolate and utilise such integrase proteins directly and experiments to this end have been carried out.

The third approach involves harnessing the enzymes responsible for the insertion of the highly mobile DNA elements known as jumping genes or transposons. These transposases act in a fairly straightforward manner, cutting a DNA segment from one place in a chromosome and pasting it in at another on the same or a different chromosome. Natural transposases isolated from a number of organisms failed to work well, so a synthetic transposase has been manufactured and the result – the Sleeping Beauty (SB) transposon – effectively enhances integration (up to about 20-fold in zebrafish embryos).

None of these approaches is yet used commercially, but work in this area is increasing knowledge of the effects of positioning on gene function. Biologists still have very little information about what most of these genes actually do. However, using model species such as the zebrafish, transgenesis of experimentally-mutated genes provides the route by which more information about the functional genomics of fish will be elucidated.

Detecting integration and expression of the transgene

The production of fish or shellfish GMOs by any method requires confirmation. We need to establish that the DNA construct is present in the GMO, that it has become integrated into the host genome and that the expected gene product is being expressed at the right level and in the right tissues for the particular commercial application for which the GMO was designed.

In order to detect the presence of the transgene in an early embryo, a simple method is to carry out dot blots. DNA is extracted from putative transgenic embryos or larvae and the extracts are spotted onto a nylon membrane in a simple grid pattern. The DNA is heat-treated to denature it and cause it to adhere to the membrane, and the membrane is then hybridised with a radiolabelled probe of a DNA sequence complementary to some part of the transgene. The membrane is then used to expose an autoradiographic film so that positive blots, where the probe has hybridised to the embryo DNA, can be identified. An alternative method is to use PCR to see if the transgene can be amplified from the putative GMO.

Dot blots and PCR help us to identify those individuals in which the transgene is present, but they do not demonstrate that the DNA has actually been incorporated into the chromosomes. In fact, because only a tiny proportion of the millions of transgene copies used per egg are ever integrated into the genome, there is bound to be extensive non-incorporated transgene. In order to provide evidence of transgene incorporation, a method involving restriction enzyme incubation and Southern blotting (Chapter 2, Box 2.3) has been used. The extracted DNA from the putative transgenic organism is incubated with a selected suite of restriction endonucleases that cut the DNA into fragments of different lengths. These fragments are then size-sorted on an electrophoretic gel and Southern blotted, with the resulting membrane being probed and visualised in the usual way. If the transgene has not become incorporated into the chromosomes then its length (and therefore position on the gel) should be the same as before it was introduced into the organism. However, where the transgene has become incorporated into the genome, it will occur in fragments of greater length than the original transgene. Not only can those individuals which have the transgene be identified using this method, but also integration of the transgene into the chromosomal material can be indicated. Unfortunately, there remains room for error here because of the vast numbers of unintegrated transgenes, many of which can recombine in unknown ways. Recombined transgenes will be longer than single transgenes and could lead to individuals, in which the transgene has not become incorporated, being wrongly identified as GMOs following blotting. An alternative approach is to use fluorescent *in situ* hybridisation (FISH, Chapter 4, Box 4.4) of a probe to the transgene to confirm its presence on chromosome spreads.

Ultimately the surest method of confirming the integration of a transgene into an individual is to demonstrate that it has passed the transgene on to its offspring in a true Mendelian fashion. Using the simplest assumption of the transgene being integrated into a single chromosome, then only half of all gametes produced by the transgenic individual (the F0 generation) will contain the transgene. Therefore the expectation is that only at most 50% of the F1 generation will contain a copy of the transgene.

How can we discover if the transgene is being expressed? We could look for mRNA produced by the gene, but it would be better to evaluate the level of production of the protein itself. One method of identifying the protein is to use an immunological approach (Box 2.11) where the presence of a particular protein can be identified by raising antibodies against it. Alternatively, of course, we can simply measure the phenotypic consequences of the transgene in the GMO. For example, with the use of an antifreeze protein, GMOs can be tested for cold tolerance. Similarly, if a growth hormone gene has been used, we can assess whether the transgenic group grow significantly faster than non-GMO controls (Table 7.2).

Table 7.2 Increase in growth rates in transgenic fish relative to controls over a single generation when using growth hormone genes from different sources

Species	Gene	% growth increase
Carp	Trout GH	50–90%
Tilapia	Tilapia GH	50–90%
	Salmon GH + ocean pout anti-freeze promoter	200–400%
Salmon	Chinook salmon GH + ocean pout anti-freeze promoter	>1000%

GH = growth hormone.

Here we will consider the results of an experiment by R.H. Devlin and colleagues on genetically modified coho salmon (*Oncorhynchus kisutch*). First, coho salmon eggs were microinjected with an 'all fish' gene construct consisting of the chinook salmon growth hormone gene and the ocean pout antifreeze protein promoter. Some of the transgenic salmon produced were male and matured at 2 years old. Five of these fish were mated to wild-type coho salmon with the expectation that about 50% of the offspring should contain the transgene. Identification was based on the PCR amplification of a specific segment of the ocean pout promoter sequence which produced a fragment only when the DNA analysed came from a transgenic fish. It was noted that the post-larval fish – alevins – were of two different colours. The majority had the normal brown coloration but some were much paler with a distinct greenish hue. PCR tests showed that the green alevins were transgenic, while the brown ones were not (Fig. 7.2). This unexpected phenotypic marker of transgenic coho salmon alevins provided a useful identifier without having to carry out PCR analysis. The greenish coloration was not a permanent marker because after first feeding, the colour of transgenics, although paler, gradually became much less distinguishable from the colour of non-transgenic fish. Analysis of the offspring from the five crosses showed that the percentages of transgenic offspring were always much less than 50%, ranging from zero to 19%. This is a common finding in F1 transgenics and probably indicates that the F0 generation were mosaics with the mosaicism extending into the germ cells. Either that, or the transgene may have been subject to loss during meiosis because of being in the haploid rather than the diploid state.

At the pre-feeding alevin stage the transgenic fish were significantly heavier and longer than the non-transgenics (Fig. 7.3), providing evidence for the enhanced activity of growth hormone in the fish. However, associated with the enhanced size was evidence of morphological abnormality, even at this early stage. Transgenic fish had larger heads than non-transgenics and this abnormality became more pronounced as the F1 fish got older (Fig. 7.4). Some severely abnormal fish had difficulty feeding due to excessive jaw growth and had problems ventilating their gills because of the extent of operculum overgrowth. Many of these F1 fish died. Similar abnormalities had been apparent in the F0 generation of transgenics.

The results of this early experiment illustrate some of the potential problems arising from transgenesis, though since these trials there have been considerable

Fig. 7.2 Agarose gels showing results of attempts to PCR amplify a portion of a DNA construct (ocean pout antifreeze promoter + chinook salmon growth hormone) in alevins of putatively transgenic coho salmon. Size markers are provided at both sides of the gels. On each gel, the two samples nearest the right-hand size markers are negative and positive control reactions. The transgene is present in the green alevins and absent in the brown alevins. (Reprinted from Devlin, R.H., Yesaki, T.Y., Donaldson, E.D. & Hew, C.-L. (1995) Transmission and phenotypic effects of an antfreeze/GH gene construct in coho salmon (*Oncorhynchus kisutch*). *Aquaculture*, **137**, 161–169. Copyright 1995, with permission from Elsevier Science.)

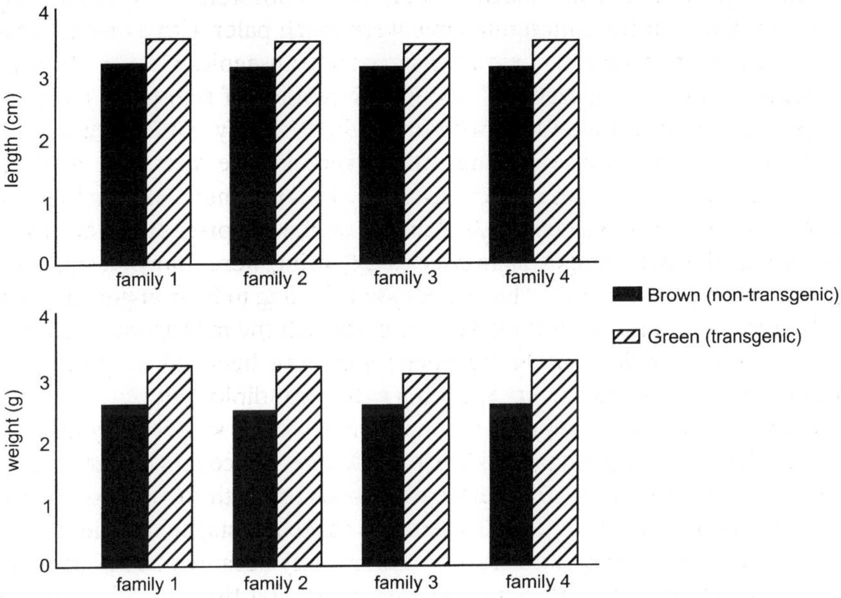

Fig. 7.3 Effect of transgenesis with a growth hormone gene on mean length (cm) and weight (g) of coho salmon alevins from four families. In every family the transgenics are significantly longer and heavier than the non-transgenics. (Modified from Devlin, R.H., Yesaki, T. Y., Donaldson, E.D. & Hew, C-L. (1995) Transmission and phenotypic effects of an antfreeze/GH gene construct in coho salmon (*Oncorhynchus kisutch*). *Aquaculture*, **137**, 161–169. Copyright 1995, with permission from Elsevier Science.)

Fig. 7.4 A one-year-old transgenic fish (length 36 cm) showing morphological abnormality of the head and an overgrown operculum. (Reprinted from Devlin, R.H., Yesaki, T.Y., Donaldson, E.D. & Hew, C.-L. (1995) Transmission and phenotypic effects of an antfreeze/GH gene construct in coho salmon (*Oncorhynchus kisutch*). *Aquaculture*, **137**, 161–169. Copyright 1995, with permission from Elsevier Science.)

improvements that reduce this type of growth abnormality in transgenic fish. Nevertheless, much more research is clearly required before transgenic aquacultured organisms can be expected to receive consumer acceptance, and thus commercial success. Unfortunately, the potential commercial value of genetic engineering has meant that most of the near-market research has been, and will continue to be, conducted by private companies that employ their own researchers and have no need to publish their findings in peer-reviewed journals. Concern and suspicion about the integrity of such research activity will inevitably arise because it is paid for by the companies themselves.

So much for transgenics – what about cloning?

Dolly, the famous first cloned sheep, was produced by transferring the nucleus from a mammary gland cell (hence the name, after Dolly Parton) of an adult sheep into an enucleated oocyte from a different sheep. Actually, of course, Dolly is not a true genetic clone but rather a chimeric animal with the mitochondrial DNA of one individual and the nuclear DNA of another, but that is hardly the point.

Nuclear transfers have been undertaken in some fish species but generally with limited success, and it is not clear precisely what purpose can be achieved by cloning fish or other aquaculture species. The aquaculture industry already has a range of methodologies available to it based around traditional genetics, ploidy manipulation and gene transfer. Indeed, true clones, with identical mitochondrial and nuclear genomes, have already been produced through two generations of gynogenesis (Chapter 6).

Genethics

Genetic engineering and cloning have opened a can of ethical worms that have wriggled into all aspects of modern society. Many of the questions raised by these technologies are of a philosophical or ethical nature. These questions are quite properly the remit of all in society – individual scientists are as entitled to their own views on these questions as is any other member of the public. The fact that they are scientists does not give them any advantage on such philosophical matters or ethical questions. However, it is entirely appropriate to ask scientists to assist the non-scientific public to understand the actual methods of genetic engineering and it is hoped that this book might in some small way contribute to this understanding.

On the other hand, there are some important questions that scientists must ask about the research they are doing in the field of genetic engineering. First we must consider what is being classed as a GMO. Is a tetraploid oyster a GMO? Such an organism is artificially engineered (indeed, even patented) and not present in the wild (as far as we know) and in that sense is unnatural. However, there is no genetic material from another species present in its genome, only two extra sets of its own chromosomes. In the view of most geneticists, tetraploids, and other products of ploidy manipulation such as triploids, androgens and gynogens, should not therefore be classed as GMOs.

How about the artificial hybridisation of two species of closely related fish that nevertheless are geographically isolated from one another? In the normal course of events (on the human time-scale) these two fish species would never meet. Here, in such hybrids, we do have genetic material from two different species in the same organism: surely these are genetically modified organisms? Well, again, most geneticists would argue that this kind of event has been happening over evolutionary time in a natural way and that we may just be speeding-up a natural process as indeed is being done in ordinary selective breeding. So such hybrids are not GMOs.

Finally, if we consider a fish that carries in its genome an extra growth gene from another vertebrate species together with a promoter from a third species, then clearly this is what we mean by a genetically modified organism. Although such a transfer of genetic material from one species to a totally unrelated one would seem at first sight to be unnatural, it is nevertheless known that in the wild, pieces of genetic material have been, and are still being, transferred occasionally between phylogenetically distinct organisms. Surprisingly, this is a natural process.

The most important practical questions to do with genetic engineering are questions about risk. Do we really know enough about what we are doing for it to be safe? We saw with the example of the change in colour of transgenic coho salmon alevins that not all effects of genetic modification are predictable. We do not yet understand many details of how the genome works and current technologies could be described as the equivalent of trying to reprogram a supercomputer with a Swiss army knife. What is the risk of the escape of genetically modified organisms into the wild? If, in spite of all efforts to reduce the chance of escape, an accident occurs and some GMOs – or their gametes – escape, what risk does that pose to (a) humans, (b) non-GMOs of the same species and (c) the environment? The balance of environmental risk against the actual value to human society has to be addressed. These questions must all be thoroughly considered before research is carried out and most developed countries have regulatory bodies that ensure that this happens.

It has been argued by some geneticists that because GMOs have extreme phenotypes, this is likely to make them less fit than wild individuals. The theory is that the added gene, although improving some market-relevant trait, acts like a deleterious allele at a locus and would be driven to extinction by natural selection. There is some evidence for this proposal, but, of course, we cannot rely on this to be the case with every individual GMO and every modified trait that is produced.

In conclusion, there is the 'precautionary principle' which can be brought to bear on the problem. It would be a sensible precaution to thoroughly research all potentially dangerous aspects of the development of genetically modified fish and shellfish before rushing ahead with commercial development. Where experimental testing of risks of escape is possible, it should be done. Some would argue that no amount of such testing will prove much, one way or another, but most observers would recognise that genetic engineering, for all its risks, is here to stay. If that is the case then careful and intelligent research in genetic engineering in aquaculture species based on the precautionary principle and with open, peer-reviewed publication, is the best way forward.

Naturally, the *raison d'être* of a company is to make a profit for its owners or shareholders. In the developed world, consumer resistance to genetically modified foods is considerable and if the company were to sell GMOs that created a problem for the aquaculturist or for the consumer then that would be bad for their business. If these issues are to be overcome, companies must have their products properly and openly tested and approved for use by government agencies. On the other hand, because of the potential earnings from new or improved genetic engineering techniques, commercial secrecy will remain a part of the balance in future developments.

Further reading

Benzie, J.A.H. (ed.) (2002) *Genetics in Aquaculture VII*. Elsevier, Amsterdam.

Devlin, R.H., Yesaki, T.Y., Donaldson, E.D. & Hew, C.-L. (1995) Transmission and phenotypic effects of an antfreeze/GH gene construct in coho salmon (*Oncorhynchus kisutch*). *Aquaculture*, **137**, 161–169.

Doyle, R.W., Herbinger, C.M., Ball, M. & Gall, G.A.E. (eds) (1996) *Genetics in Aquaculture V*. Elsevier, Amsterdam.

Fingerman, M. & Nagabhushanam R. (eds) (2000) *Recent Advances in Marine Biotechnology, Volume 4, Aquaculture Part B, Fishes*. Science Publishers Inc., Enfield, NH.

McAndrew, B. & Penman, D. (eds) (1999) *Genetics in Aquaculture VI*. Elsevier, Amsterdam.

Glossary

acrocentric (telocentric) Descriptive of a chromosome which has the centromere very close to one end.

additive genetic variance That part of the genetic variance in a quantitative trait which is due to the presence or absence of particular alleles.

adenine One of the bases acting as a coding letter in nucleotides in DNA and RNA molecules; symbol A.

AIS *See* alike in state.

alike in state Descriptive of alleles present in a homozygote that are alike but which are not recent copies of an ancestral allele.

allele One of two or more alternative forms of a DNA sequence at a locus.

allozymes Different forms of an enzyme produced by genetic variation at the locus coding for that enzyme.

amino acids Chemicals which form the basic building blocks of proteins.

anadromous Descriptive of a species which spawns and reproduces in fresh water but spends some or much of its adult life at sea.

anaphase The phase of mitosis or meiosis during which chromosomes migrate to opposite poles of the spindle and separate to daughter cells.

androgen In genetics, an individual whose chromosomes come only from its male parent (alternatively, a masculinising hormone).

aneuploid Possessing a number of chromosomes which is not an exact multiple of the haploid chromosome number.

annealing The pairing up of single strands of DNA which have complementary sequences. In the case of PCR, this is where the primers join to their complementary sequences on the template DNA.

antibodies Immunoglobulin molecules which are produced in the blood of vertebrates in response to foreign proteins.

antigenic determinants Sites on the exterior of proteins which initiate the production of antibodies in the blood of vertebrates.

antiserum Serum (blood with the cells removed) which contains antibodies to a specified protein.

autoradiography The process of exposing a photographic negative to identify the position of very small quantities of radioactively labelled DNA.

autosomes Chromosomes which are not sex chromosomes.

bacteriophage (phage) A virus which infects bacteria.

base Part of a nucleotide in DNA or RNA. Chemically, adenine (A) and guanine (G) are purines while cytosine (C), thymine (T) (in DNA) and uracil (U) (in RNA) are pyrimidines. The sequence of the bases along the DNA or RNA molecules provides the coding instructions for proteins.

bioinformatics The development and use of software to study gene and amino acid sequence data.

biolistics A method for inserting DNA fragments into cells by bombarding them with minute gold particles covered with the DNA.

bivalents Pairs of homologous chromosomes joined together during meiosis.

breeding value The value given to an individual which indicates how much improvement over the population mean of a trait that its offspring are likely to produce.

broad-sense heritability The proportion of phenotypic variance in a quantitative trait which can be attributed to all kinds of genetic variation whether this is directly heritable or not. It includes additive, dominance and epistatic genetic variance.

cDNA *See* complementary DNA

central dogma The understanding that information is transferred from DNA to RNA to protein, believed to be broken only by the retroviruses.

centromere The constricted region on the chromosome where it attaches to the spindle during cell division and which divides the chromosome into distinct arms.

chiasma (pl. chiasmata) The point on the chromosome at which recombination takes place.

chloroplast Organelle present in plant cells in which photosynthesis takes place.

chromatid A daughter chromosome still joined to its pair at the centromere, visible during meiosis or mitosis.

chromosomes Structures present in the cell nucleus which contain the DNA in combination with proteins.

cloning A general term for the process of isolating and replicating fragments of DNA by inserting them into vectors (such as plasmids), integrating the vectors into hosts (such as the bacterium *E. coli*) and allowing the host and vector to replicate. Alternatively, the process of artificially creating a number of genetically identical individuals.

coding strand (non-template strand, sense strand or + strand) The strand of DNA which carries the sequence coding for the gene. This is the sequence that the mRNA carries.

co-dominant Descriptive of loci where both alleles can be identified. Compare dominant, recessive.

codon A set of three bases in the coding strand of DNA or in mRNA which specifies a particular amino acid.

coefficient of genetic diversity (G_{ST}) An index of the degree to which a group of populations is genetically subdivided.

complementary base pairing The arrangement whereby each base pairs up with only one other base on the opposite strand of the two-stranded DNA molecule, such that adenine always pairs with thymine and guanine always pairs with cytosine. This means that the sequence of one strand can be used to predict the sequence of the complementary strand.

complementary DNA (cDNA) DNA copied from an RNA template by the enzyme reverse transcriptase.

copy number The number of copies of a DNA fragment or plasmid.

crossing-over The process by which genetic material is exchanged between non-sister chromatids during meiosis.

cryoprotectants Chemicals used during the process of cryopreservation to reduce the formation of ice crystals.

cytosine One of the bases acting as a coding letter in nucleotides in DNA and RNA molecules; symbol C.

dam Mother.

ddNTPs (dideoxynucleotide triphosphates) Modified versions of deoxynucleotide triphosphates (dNTPs) used as chain terminators in DNA sequencing.

deletion Loss of part of a chromosome.

denaturation The separation of the two strands of the DNA molecule.

dendrogram An illustration in the form of a tree, used to show genetic relatedness between different populations or taxa.

deoxynucleotide triphosphates (dNTPs) The four nucleotide building blocks for DNA synthesis (dATP, dCTP, dGTP, dTTP).

deoxyribose nucleic acid (DNA) The molecule which forms the genetic code. A polymer double helix of two strands of nucleotides each consisting of a sugar (2-deoxy-D-ribose), a phosphate (PO_3) and one of four nitrogenous bases, adenine (A), cytosine (C), guanine (G) and thymine (T).

dideoxynucleotide triphosphates (ddNTPs) Modified versions of deoxynucleotide triphosphates (dNTPs) used as chain terminators in DNA sequencing.

dimer Something which consists of two parts, such as a protein consisting of two subunits.

diploid Possessing two sets of chromosomes.

DNA *See* deoxyribose nucleic acid.

DNA construct The DNA transferred during transgenesis, consisting of a transgene spliced to its promoter.

DNA library Genomic DNA from a particular organism fragmented and cloned into a vector/host system (such as the plasmid/*E. coli* system).

DNA ligase The enzyme which catalyses the joining together of fragments of DNA.

dNTPs (deoxynucleotide triphosphates) The four nucleotide building blocks for DNA synthesis (dATP, dCTP, dGTP, dTTP).

dominance genetic variance That part of the genetic variance in a quantitative trait which is due to the presence or absence of particular genotypes.

dominant Descriptive of an allele which solely determines the phenotype of an individual regardless of whether it is present in the homozygous or heterozygous condition. The alternative allele is the recessive allele.

dot blots A modification of Southern blotting in which DNA samples are spotted in a grid pattern onto a nylon membrane and probed to identify those containing a particular sequence.

duplication An extra copy of a gene or a chromosomal segment which occurs on the same or a different chromosome.

effective population size (N_e) The number of individuals in the population which contribute genetically to the next generation.

electroporation A method for inserting DNA into cells which involves applying a short electrical pulse to the cells to increase their porosity.

epistatic genetic variance That part of the genetic variance in a quantitative trait which is due to the interaction between genotypes at different loci. Sometimes called non-allelic interaction genetic variance.

euchromatic Descriptive of relatively uncondensed regions of the chromosome thought to contain mainly coding DNA.

eukaryote An organism containing cells with a membrane-bound nucleus.

exon That part of the sequence of a gene which is translated into an amino acid sequence.

family selection In quantitative genetics, selection of particular individuals for mating because they belong to a family which has a high value for a quantitative trait.

first polar body The smaller of the two daughter cells resulting from the meiosis I division of the primary oocyte.

F_{IS} A statistic used to estimate deviations from the Hardy–Weinberg model within subpopulations (which make up the total population).

F_{IT} A statistic used to estimate deviations from the Hardy–Weinberg model within the total population (which is made up of subpopulations).

F_{ST} A statistic used to estimate the degree to which a total population is genetically divided into subpopulations.

***F*-statistics** A set of statistics (F_{IS}, F_{IT} and F_{ST}) used to estimate deviations from the Hardy–Weinberg model in populations and to estimate the degree to which a group of populations is genetically subdivided.

gametogenesis The process of gamete formation.

G-banding Chromosomal banding pattern produced using Giemsa stain.

gene A unit of information corresponding to a discrete segment of DNA that codes for the sequence of an RNA molecule or for the amino acid sequence of a protein.

gene expression The process of transcription and translation of the DNA code into a protein.

general combining ability The mean performance of cross-bred offspring of a particular line in relation to the average performance of cross-bred offspring from all possible lines.

genethics The field of ethics in relation to genetic modification and genetic engineering.

genetically modified organism (GMO) An organism whose genome has been artificially modified by the addition of genetic material from another species.

genotype The genetic composition of an individual at one or many loci.

GMO (genetically modified organism) An organism whose genome has been artificially modified by the addition of genetic material from another species.

G_{ST} Coefficient of genetic diversity, an index of the degree to which a group of populations is genetically subdivided.

guanine One of the bases acting as a coding letter in nucleotides in DNA and RNA molecules; symbol G.

gynogen An individual which contains only chromosomes from its female parent. Haploid gynogens of fish and shellfish are non-viable but viable diploid gynogens can be produced.

h^2 (heritability) The proportion of phenotypic variance in a quantitative trait which can be attributed to genetic variation.

half-siblings Individuals related through having either the same mother or the same father.

haploid Possessing a single copy of each chromosome.

haplotype A set of alleles inherited together, such as the mitochondrial DNA molecule.

Hardy–Weinberg model The model which relates genotype frequency to allele frequency at a locus in a population.

heritability (h^2) The proportion of phenotypic variance in a quantitative trait which can be attributed to genetic variation.

heterochromatic Descriptive of very condensed regions of the chromosome thought to consist mainly of non-coding DNA.

heterozygous Coding for two different alleles at a locus.

homologous pair A pair of chromosomes in diploids, one of which is derived from the male parent, the other from the female parent. Chromosomes become arranged together as homologous pairs during meiosis.

homozygous Coding for two identical alleles at a locus.

hybrid vigour Phenomenon of the hybrid between two lines or species exhibiting greater fitness than either parent group.

IBD *See* identical by descent.

identical by descent (IBD) Descriptive of alleles present in a homozygote which have arisen by replication of a single allele from a direct ancestor

inbreeding Mating between close relatives.

inbreeding coefficient An index which estimates the level of deviation from the Hardy–Weinberg model caused by inbreeding.

inbreeding depression The reduction in fitness of inbred offspring relative to outbred.

independent assortment A process of chromosome shuffling whereby each homologous pair of chromosomes is divided between the two daughter cells during meiosis I independently of the other homologous pairs.

independent culling Selective regime which involves breeding only from individuals which demonstrate high values for each of two or more traits.

individual selection Artificial selection involving breeding from those individuals scoring highest for a particular trait. Also called mass selection.

integrase protein A viral protein which assists with the integration of viral DNA into chromosomes.

interploid triploid A triploid produced by mating a diploid with a tetraploid.

intron That part of the sequence of a gene which is not translated into an amino acid sequence, spliced out of pre-messenger RNA.

inversion Reversal of the position of a fragment of a chromosome.

karyotype Representation of the complete chromosome complement of an organism, showing the position of the centromere on each chromosome pair.

LINE (long interspersed nuclear elements) Identical or very similar DNA sequences thousands of base pairs long occurring in multiple copies dispersed around the genome.

linkage The physical association between two genes that are on the same chromosome which means that alleles at each tend to be inherited together.

linkage analysis The method whereby recombination frequencies are used to estimate the relative positions of loci on a chromosome.

linkage disequilibrium Situation where alleles at two linked loci on a chromosome are non-randomly associated with each other. The loci may be linked by physical proximity or by selective forces.

linkage equilibrium Situation where alleles at two loci are inherited randomly with respect to each other because the loci are on different chromosomes and are not co-selected.

linkage group The genes carried on any one chromosome.

lipofection A method for inserting DNA into cells by allowing synthetic lipid vesicles containing encapsulated DNA to be taken up directly.

locus (pl. loci) The site on the chromosome at which a gene or genetic marker is located.

long interspersed nuclear elements (LINE) Identical or very similar DNA sequences thousands of base pairs long occurring in multiple copies dispersed around the genome.

long terminal repeats (LTR) Identical or very similar DNA sequences with long repetitive elements at the ends occurring in multiple copies dispersed around the genome.

LTR *See* long terminal repeats.

mass selection Artificial selection involving breeding from those individuals scoring highest for a particular trait. Also called individual selection.

meiogyne A diploid gynogen produced by suppression of cell division following either meiosis I or meiosis II in an egg activated by a sterizoon.

meiosis The process of cell division in which the chromosomes in diploid germ cells are replicated and passed to haploid gametes.

meiosis I The first division in the process of meiosis, in which the homologous pairs of chromosomes are divided.

meiosis II The second division in the process of meiosis, in which sister chromatids are separated before cell division so that parent and daughter cells have the same chromosome number.

messenger RNA (mRNA) The transcript of a protein coding gene.

metacentric Descriptive of a chromosome which has an approximately central centromere.

metaphase The phase of mitosis or meiosis when chromosomes are aligned on the equator of the spindle prior to separation.

methylation Replacement of hydrogen by a methyl group, CH_3, in a DNA base.

microinjection A method for inserting DNA into cells by injection.

microsatellites Variable number tandem repeats (VNTR) where the repeat motif is from one to four base pairs in length.

minisatellites Variable number tandem repeats (VNTR) where the repeat motif is from five to 100 base pairs in length.

mitochondrial DNA (mtDNA) A circular molecule of DNA present in the mitochondria of eukaryotic cells.

mitochondrion (pl. mitochondria) An energy-generating organelle present in eukaryotic cells.

mitogyne A diploid gynogen produced by suppression of the first cleavage division in an egg activated by a sterizoon.

mitosis The process of cell division in which the chromosomes are replicated and passed to daughter cells.

molecular clock A theoretical clock based on the rates at which point mutations or amino acid substitutions become fixed in particular evolutionary lineages. Assuming these rates are approximately constant, the clock enables us to use sequence differences to estimate the time since taxa diverged.

monomer Something which consists of a single part, such as a protein consisting of a single subunit.

mosaic Descriptive of an organism containing cells of different ploidies, or containing some cells with a transgene and others without.

mRNA (messenger RNA) The transcript of a protein coding gene.

mtDNA (mitochondrial DNA) A circular molecule of DNA present in the mitochondria of eukaryotic cells.

narrow-sense heritability The proportion of phenotypic variance in a quantitative trait which can be attributed to additive genetic variance and which is therefore amenable to simple artificial selection.

N_e (effective population size) The number of individuals in the population which contribute genetically to the next generation.

NLS (nuclear-localising sequences) Sequences of amino acids which assist molecules to get into the cell nucleus.

non-allelic interaction genetic variance That part of the genetic variance in a quantitative trait which is due to the interaction between genotypes at different loci. Also called epistatic genetic variance.

non-sister chromatids Chromatids in opposite chromosomes of a homologous pair.

non-template strand (sense, + or coding strand) The strand of DNA which carries the sequence coding for the gene. This is the sequence that the mRNA carries.

nuclear-localising sequences (NLS) Sequences of amino acids which assist molecules to get into the cell nucleus.

nucleotide A unit of a nucleic acid consisting of a sugar (deoxyribose in DNA or ribose in RNA), a phosphate (PO_3) and one of the bases, adenine (A), cytosine (C), guanine (G) and thymine (T) in DNA or uracil (U) in RNA.

null alleles Alleles which cannot be detected by the method used to detect other alleles at that locus.

oligomer Something which consists of a number of parts, such as a protein consisting of several subunits.

operational taxonomic unit (OTU) The taxonomic unit used in the preparation of a dendrogram.

OTU *See* operational taxonomic unit.

ovum (pl. ova) The single haploid cell resulting from meiosis of a female germ cell – the female gamete, the egg.

paracentric inversion Inversion of part of a chromosome which does not include the centromere.

PCR (polymerase chain reaction) A method for making millions of copies of short (0.1–4.0 kb) fragments of DNA.

pericentric inversion A chromosomal inversion which spans the centromere.

phage (bacteriophage) A virus which infects bacteria.

plasmid Small self-replicating circular DNA molecules present in bacteria and yeasts. They are used in genetic engineering as vectors to insert foreign DNA into bacteria for cloning.

pleiotropy The condition where a single gene influences more than one phenotypic character or trait.

point mutation A change in the identity of a base in a DNA sequence.

polygenic trait A trait controlled by many loci. Also known as a quantitative trait.

polymerase chain reaction (PCR) A method for making millions of copies of short (0.1–4.0 kb) fragments of DNA.

polynucleotide chain A chain of nucleotides making up a strand of DNA or RNA.

polyploid Possessing more than two complete sets of chromosomes.

position effect The effect of chromosomal positioning on the potential expression of a transgene in a GMO. Transgenes which locate in heterochromatic regions are seldom switched on.

primary oocyte A female germ cell which has begun the first meiotic division.

prime Symbol (′) assigned to the numbered carbon atoms in the deoxyribose or ribose sugar in DNA or RNA respectively.

primers Short oligonucleotides, manufactured to a specified sequence, which are used in the polymerase chain reaction process.

promoter A DNA sequence which is involved in the initiation of transcription and translation of a gene.

prophase The first phase of meiosis or mitosis.

pseudogene A sequence of DNA which is non functional but which shows significant sequence homology to a known functional gene.

purines Chemical description of the nitrogenous bases adenine and guanine present in DNA and RNA. Purines will not bond with one another but will bond with pyrimidines.

pyrimidines Chemical description of the nitrogenous bases cytosine and thymine present in DNA and of uracil in RNA. Pyrimidines will not bond with one another but will bond with purines.

QTL (quantitative trait loci) Loci which have some effect on the expression of a quantitative trait.

qualitative trait A trait defined by simple discrete categories, often controlled by only one or a few genes.

quantitative genetics The study of the inheritance of quantitative traits.

quantitative trait A trait which shows a continuous graded variation around a mean and can be measured quantitatively. Such traits, also called polygenic traits, are usually partly controlled by many loci and partly controlled by the environment.

quantitative trait loci (QTL) Loci which have some effect on the expression of a quantitative trait.

quaternary structure The number of subunits present in a protein.

random genetic drift The process of allele frequencies changing from one generation to the next as a result of sampling effects.

realised heritability (realised h^2) The broad sense heritability actually achieved following selection.

recessive Descriptive of an allele which is only expressed in the homozygous state.

recombinant DNA DNA produced by joining DNA molecules from two different sources.

recombination The process of genetic material being exchanged between non-sister chromatids during meiosis.

recombination frequency The proportion of gametes which are the result of recombination events between two specified loci during meiosis. Recombination frequencies are used to estimate distances separating loci on a chromosome.

recombination hotspot Position on a chromosome where recombination events are more frequent than expected. Recombination frequencies for loci close to recombination hotspots are biased estimators of the position of those loci in a linkage group.

refugia Regions or habitats in which populations or entire species are able to survive large-scale environmental changes such as glaciations.

replication origins The points at which the two strands of the DNA molecule open up to allow the process of replication to occur.

replication slippage The process by which the number of repeat motifs in microsatellites is thought to increase or decrease during DNA replication.

reporter gene Genes used in experimental transgenesis which have an obvious effect on the embryo, enabling different incorporation methods to be easily assessed.

response to selection In quantitative genetics, the difference between the mean value for the trait in the overall parental population and the mean value of the trait among the offspring of the selected parents.

restriction endonucleases Enzymes extracted from bacteria which cleave DNA at specific base sequences.

restriction mapping The process of identifying the positions of restriction sites for particular restriction endonucleases on chromosomes or mitochondrial DNA molecules.

reverse transcriptase Enzyme which synthesises DNA from an RNA template. Such DNA is called complementary DNA or cDNA.

ribose nucleic acid (RNA) A single-stranded form of nucleic acid involved in the transcription and translation of the DNA code, consisting of a polymer of nucleotides each composed of a sugar (ribose), a phosphate (PO_3) and one of four nitrogenous bases, adenine (A), cytosine (C), guanine (G) and uracil (U).

ribosomal RNA (rRNA) RNA molecules present in ribosomes, involved in the translation of the genetic code from messenger RNA into a sequence of amino acids.

RNA *See* ribose nucleic acid.

rRNA *See* ribosomal RNA.

satellites Regions in the genome which contain tandem repeats of DNA sequences. The number of repeated sequences can be variable between individuals and these types of marker are also called variable number of tandem repeats (VNTRs).

secondary oocyte The largest of the two daughter cells resulting from the meiosis I division of the primary oocyte.

second polar body The smaller of the two daughter cells resulting from the meiosis II division of the secondary oocyte.

selection The process of one allele or genotype at a locus having a different fitness to other alleles or genotypes such that an organism carrying that allele or genotype has a greater or lesser chance of reproducing.

selection differential In quantitative genetics, the difference between the mean value for the trait in the population and the mean value for the trait in individuals selected for breeding.

semi-conservative replication Description of the process of replication of DNA where each copied molecule contains one strand derived from the parent molecule and one newly synthesised strand.

sense strand (non-template strand, + strand, coding strand) The strand of DNA which carries the sequence coding for the gene. This is the sequence that the mRNA carries.

sex chromosomes A pair of chromosomes which control the sex of the individual, for example female XX, male XY.

short interspersed nuclear elements (SINE) Identical or very similar DNA sequences hundreds of base pairs long occurring in multiple copies dispersed around the genome.

siblings Individuals having the same mother and the same father.

SINE *See* short interspersed nuclear elements.

sire Father.

sister chromatids Chromatids from the same chromosome.

somatic Descriptive of the non-reproductive parts of an organism.

Southern blotting The process of transferring DNA from a gel onto a nylon membrane in order to create a stable base on which to probe for particular DNA sequences.

specific combining ability The mean performance of cross-bred offspring of a particular line when crossed with another individual line.

spindle A structure present in the cytoplasm of the cell which guides the movement of chromosomes during cell division.

sterizoon (pl. sterizoa) A spermatozoon which has been irradiated to denature the chromosomes but which is still motile and capable of activating an egg. Sterizoa are used in gynogenesis.

strand A polynucleotide chain making up half of a DNA molecule or all of an RNA molecule. In the DNA molecule the two strands are oriented as a double helix.

submetacentric Descriptive of chromosomes which have the centromere between the centre and one end.

subunits Units making up the quaternary structure of a protein.

syngamy The joining of the haploid male and female pronuclei to produce a diploid zygote.

tandem selection Selection for one trait in one generation and for an alternative trait in the next generation.

taxon (pl. taxa) A grouping of organisms assigned to a particular category of classification, e.g. a species, genus or order.

telocentric (acrocentric) Descriptive of a chromosome which has the centromere at one end.

telomere The end of a chromosome, usually consisting of a region of many short repeats.

template DNA The DNA which is copied, particularly in the polymerase chain reaction method.

template strand The strand of DNA which carries the sequence complementary to that coding for the gene, acting as the template against which mRNA is synthesised.

test cross An experimental cross between two individuals, one of which is homozygous at two specified loci and the other of which is heterozygous at the two loci. The F1 progeny genotypes from a test cross allow direct estimation of recombination frequencies between the two loci.

tetramer Something which consists of four parts, such as a protein consisting of four subunits.

tetraploid Containing four sets of chromosomes.

the central dogma The understanding that information is transferred from DNA to RNA to protein, believed to be broken only by the retroviruses.

thymine One of the bases acting as a coding letter in nucleotides in the DNA molecule; symbol T.

trait A quantifiable phenotypic character.

transcription The process whereby the genetic code in DNA is copied into an RNA molecule.

transfer RNA (tRNA) Short RNA molecules involved in translating codons in messenger RNA into amino acids.

transformed cell A bacterial cell into which recombinant DNA has been inserted.

transgene Gene which is inserted into the genome of another species during transgenesis.

transgenesis The creation of an organism which contains a gene (a transgene) from another species.

translation The synthesis, involving tRNA and rRNA, of a chain of amino acids according to a particular mRNA sequence.

translocation Movement of a fragment from one chromosome to another chromosome.

transposases Enzymes which assist transposons to move from one position to another in the genome.

transposons Highly mobile DNA elements that move from one position in genome to another, also known as jumping genes.

triploid Containing three sets of chromosomes.

tRNA (transfer RNA) Short RNA molecules involved in translating codons in messenger RNA into amino acids.

uracil One of the bases acting as a coding letter in nucleotides in the RNA molecule; symbol U.

variance The condition of having variation around a mean. In statistical terms variance is calculated as the mean of the squares of individual deviations from the mean.

within family selection In quantitative genetics, artificial selection for a particular trait by breeding from individuals which are highly ranked for that trait within their family.

zygote The diploid cell produced by the joining together of the haploid male and female pronuclei during syngamy.

Index